Emergency Information

Name of Chemist ____________________

Address (optional) ____________________

Telephone Number (optional) ____________________

Laboratory Information

Course Number ____________________

Lab Section Number ____________________

Instructor's Name ____________________

Laboratory Building and Room Number ____________________

Desk Number ____________________

Stock Room Number ____________________

Location of Safety Equipment Nearest to Your Laboratory Bench

Main Power Shut-Off Switch ____________________

Main Gas Shut-Off Valve ____________________

Telephone ____________________

First Aid Kit ____________________

Fire Extinguisher ____________________

Fire Blanket ____________________

Safety Shower ____________________

Eye Wash Fountain ____________________

Campus Medical Office ____________________

Local Emergency Telephone Numbers

Campus Safety ____________________

Ambulance ____________________

Fire Department ____________________

Police ____________________

AF718833

Laboratory Experiments

David B. Macaulay William Rainey Harper College
Joseph M. Bauer William Rainey Harper College
Molly M. Bloomfield Oregon State University

General, Organic, and Biological Chemistry: An Integrated Approach

BICENTENNIAL 1807 WILEY 2007 BICENTENNIAL

John Wiley & Sons, Inc.

Cover Photo: © Cris Benton, Kite Aerial Photography

Copyright © 2007 John Wiley & Sons, Inc. All rights reserved. No part of this publication may be reproduced, stored in a retrieval system or transmitted in any form or by any means, electronic, mechanical, photocopying, recording, scanning, or otherwise, except as permitted under Sections 107 or 108 of the 1976 United States Copyright Act, without either the prior written permission of the Publisher, or authorization through payment of the appropriate per-copy fee to the Copyright Clearance Center, Inc., 222 Rosewood Drive, Danvers, MA 01923, (978)750-8400, fax (978)750-4470 or on the web at www.copyright.com. Requests to the Publisher for permission should be addressed to the Permissions Department, John Wiley & Sons, Inc., 111 River Street, Hoboken, NJ 07030-5774, (201)748-6011, fax (201)748-6008, or online at http://www.wiley.com/go/permissions.

To order books or for customer service please, call 1-800-CALL WILEY (225-5945).

ISBN-13 978- 0-470-040287
ISBN-10 0-470-040289

Printed in the United States of America

10 9 8 7 6 5 4 3 2

Printed and bound by Bind Rite Graphics

A Note From the Publisher

Laboratory Experiments to accompany General, Organic, and Biological Chemistry: An Integrated Approach, highlights the relevance and application of basic chemical principles to biological systems. The laboratory experiments are designed to relate to the student's personal experience and use many common household and commercial products.

This manual accompanies ***General, Organic, and Biological Chemistry: An Integrated Approach***, by Kenneth Raymond. The laboratory exercises are closely integrated with the text subject matter, with specific text references available for the majority of the experiments. The manual assumes no prior student experience in the chemistry laboratory.

ORGANIZATION OF THE MANUAL

Safety in the Laboratory

This section provides safety rules and precautions. It contains the discussion on MSDS sheets and NFPA hazard ratings.

Fundamental Laboratory Operations

This highly illustrated section demonstrates and explains the most commonly performed laboratory operations. Frequent reference is made to various portions of this section throughout the manual.

Twenty-One Laboratory Experiments

These experiments supply the instructor with a wide selection for use during a one-semester course, while also providing experiments to cover the laboratory portion of a two-semester course. Most experiments can be completed within two hours; by expanding pre-laboratory and post-laboratory discussions, a three-hour period can be utilized. Each experiment consists of the following parts: an Introduction, Procedure, Prelab Questions, a Report, and Related Questions. The **Introduction** explains the chemical principles relevant to the experiment; in many instances examples pertinent to the allied heath fields are provided. Most experiments make a reference to specific chapters in ***General, Organic, and Biological Chemistry: An Integrated Approach***. The Procedure section presents the instructions for each experiment. **Prelab Questions** are to be completed before the start of the laboratory period. Experimental data, calculations, and conclusions are to be recorded in the **Report** section, which also includes specific questions about the student's observations. Additional questions that require the student to apply some of the experimental principles to other problems are found in the **Related Questions** section of each experiment. The answers to these questions may require reference to the text or other sources such as The Merck Index or the CRC Handbook of Chemistry and Physics.

INSTRUCTOR'S MANUAL

The Instructor's Manual provides teaching hints and suggestions, expected data and results, answers to questions, a list of supplies and chemicals for each experiment, and directions for preparing solutions.

ACKNOWLEDGEMENTS

We would like to thank Kenneth Raymond for his helpful feedback during the development of this project. We would like to acknowledge Bette Kreuz for her expert analysis and careful review of these experiments. Thank you to Joan Kalkut for her careful editing and helpful suggestions. Special thanks to Wendy Perez for managing the production process flawlessly.

Table of Contents

Safety in the Laboratory

Be certain to read these instructions carefully **before attending the first laboratory session**.

BEFORE COMING TO THE LABORATORY

Before you come into the laboratory, carefully read and study the laboratory experiment to be performed. The most common reason that students do not complete an experiment in the time allowed is lack of prelaboratory preparation. Your instructor will provide additional instructions for an experiment when needed, and will let you know whether the experiment is to be performed individually, in pairs, or in groups. You may be requested to complete and hand in the Prelab Questions before performing an experiment. Record all data for each experiment on the Report as soon as you have completed a particular laboratory operation. The Related Questions may be completed after the laboratory period.

RULES FOR WORKING IN THE LABORATORY

At the Laboratory Station

1. Turn off the Bunsen burner when you are not using it.
2. Turn off hotplates (including the magnetic stirring motor, if present) or any other electrical apparatus when not in use. Unplug the apparatus when you are finished with the experiment.
3. When heating beakers or flasks, use a wire gauze under the container for more even distribution of heat and more rapid heating.
4. If any hazardous reagents are spilled, notify your instructor at once.

Handling Chemicals

1. Check odors cautiously; never taste a chemical.
2. To avoid spattering of acids which can cause burns, always add acid to water—never add water to acid.
3. Before obtaining any reagents, carefully read the labels on the bottles twice. Many chemicals have similar names.
4. Do not remove a reagent from the designated dispensing area. Use the appropriate container (test tube, beaker, etc.) for obtaining chemicals.
5. To avoid unnecessary waste, obtain only the amount of chemicals called for in an experiment. Your instructor will tell you the proper procedure for dispensing liquids and solids. Never insert medicine droppers or scoops into reagent bottles. See Fundamental Laboratory Operations, Part 1 on page 8.
6. Never return unused chemicals to the original dispensing bottle.
7. Follow the instructor's directions for disposal of chemicals. When no specific directions are given, dispose of nonhazardous, water-soluble substances in the sink, and put insoluble materials such as filter paper in the waste basket.

SAFETY PRECAUTIONS

1. Perform only the experiment assigned; do not experiment on your own.
2. Carefully note any precautions stated in the laboratory experiment, especially those written in capital letters.
3. Always wear safety goggles of the specific type required in your laboratory. Do not wear sandals or open-toed shoes. Do not wear clothes with loose, long sleeves.
4. Since many reagents are flammable or toxic, smoking, eating, chewing gum, or drinking are not permitted in the laboratory.
5. Hair can easily catch on fire; if you have long hair, tie it back.
6. Know the location of the fire extinguisher, fire blanket, eye wash, and safety shower. Your instructor will describe their use.
7. If you wear contact lenses, be especially wary of splashing a chemical in your eyes. It is possible for liquid reagents to seep under the lens so that flushing your eyes will not remove all the chemical. The lens should be removed as quickly as possible so that both the lens and eye can be thoroughly washed. Plastic contact lenses can also absorb vapors of organic solvents.
8. When noxious vapors are likely to be produced or when volatile or flammable chemicals are used, perform the laboratory operations under a fume hood.
9. Be especially conscious of hot objects and broken glassware. Burns and cuts are the most common types of injury in the laboratory. Promptly report any injury, no matter how slight, to your instructor.
10. When heating a liquid in a test tube, clamp the tube at the top, and concentrate the flame near the surface of the liquid. Do not point the open end of the test tube toward another person. Do not heat graduated cylinders or bottles or apply an open flame to a watch glass. See Fundamental Laboratory Operations, Part 6, page 19.
11. When heating liquids on hotplates, be sure to place the container on a wire gauze and add either a boiling stone or magnetic stir bar to the liquid to avoid bumping. Never add a boiling stone or stir bar to a hot liquid as this may cause extensive foaming and loss of the liquid out of the container. Do not use boiling stones when heating acidic solutions. If a magnetic stir bar cannot be used, place a glass stirring rod in the liquid to promote even boiling. On a hotplate, never heat a container to dryness. See Fundamental Laboratory Operations, page 19.
12. When using hotplates or any electrical apparatus, avoid contact between water and the apparatus or any of the electrical connections.
13. Be sure to use the proper devices for holding hot beakers, test tubes, crucibles, etc. Your instructor will demonstrate the correct procedure.
14. If you do not understand the instructions in this manual, ask your instructor for clarification before performing the procedure.

LABELING OF CHEMICALS

Every chemical in a chemistry laboratory must be properly labeled. Many chemicals have similar names and you should read the name twice. If a chemical is a solution, the concentration will also appear on the label. Solution concentration is commonly described by molarity (e.g., 6M HCl) or by percent concentration (e.g., 0.9% NaCl).

Materials are often labeled with information about the hazards of the chemicals. One of the most common systems was developed by the **National Fire Protection Association* (NFPA)**. NFPA labels describe four hazards—health, fire, reactivity, and specific hazards. For each hazard a numerical rating of 0 to 4 is given. Each hazard rating is displayed on a color-coded diamond-shaped label, divided into 4 smaller colored labels. See Figure 0-1.

---

*See *Fire Protection Guide to Hazardous Materials*, 13th edition (2001), National Fire Protection Association, Quincy, MA.

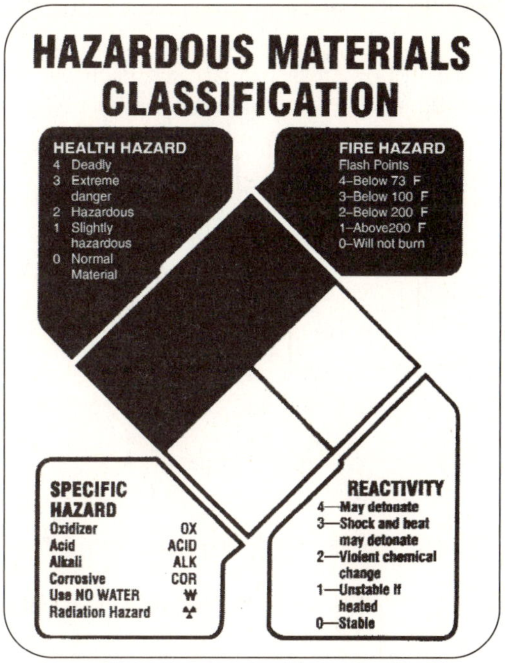

Figure 0-1 NFPA hazardous materials classification system.

NFPA Health Hazard Ratings
Color Code: BLUE

Rating	*Type of Possible Injury*
4	Materials that could cause death or major residual injury.
3	Materials that on short exposure could cause serious temporary or residual injury.
2	Materials that on intense or continued but not chronic exposure could cause temporary incapacitation or possible residual injury.
1	Materials that on exposure could cause irritation but only minor residual injury.
0	Materials that under fire conditions would offer no hazard beyond that of ordinary combustible material.

NFPA Flammability Hazard Ratings
Color Code: RED

Rating	*Susceptibility of Materials to Burning*
4	Materials that will rapidly or completely vaporize at atmospheric pressure and normal ambient temperature, or that are readily dispersed in air and that burn readily.
3	Liquids and solids that can be ignited under almost all ambient temperature conditions.
2	Materials that must be moderately heated or exposed to relatively high ambient temperature before ignition can occur.
1	Materials that must be preheated before ignition can occur.
0	Materials that will not burn.

NFPA Reactivity (Stability) Hazard Ratings
Color Code: YELLOW

Rating	*Susceptibility to Release of Energy*
4	Materials that in themselves are readily capable of detonation or of explosive decomposition or reaction at normal temperatures and pressures.
3	Materials that in themselves are capable of detonation or explosion or reaction but which require a strong initiating source or which must be heated under confinement before initiation or which react explosively with water.
2	Materials that readily undergo violent chemical change at elevated temperatures and pressures or which react violently with water or which may form explosive mixtures with water.
1	Materials that in themselves are normally stable, but which can become unstable at elevated temperatures and pressures.
0	Materials that in themselves are normally stable, even under fire exposure conditions, and which are not reactive with water.

NFPA Specific Hazards

Symbol	*Special Hazard*
~~W~~	Materials that demonstrate unusual reactivity with water.
OX	Materials that demonstrate oxidizing properties.

The NFPA system was developed for the needs of firefighters, not for chemists. The NFPA ratings represent hazard under fire conditions as opposed to normal laboratory use, a situation that may distort hazard characteristics. Many chemicals are not rated. Several chemical companies have also developed similar, more comprehensive, rating systems but thus far they are not as widely used as the NFPA system, nor do they use standardized colors and ratings.

MATERIAL SAFETY DATA SHEETS

Material Safety Data Sheets (MSDSs) are safety reports that must be supplied by the manufacturer with each chemical purchased. An MSDS lists 16 pertinent points that enumerate all the known vital information regarding the safe shipping, storage, use, handling, and disposing of a chemical. MSDS sheets, while providing specific information and first aid measures, also provide a wealth of information about their physical and chemical properties. Any institution must have MSDSs readily available for all the chemicals it uses.

Another useful source of hazard information is **The Merck Index**. This one-volume encyclopedia of chemicals is published by Merck & Co., Inc., Rahway, NJ. It contains more than 10,000 chemicals and lists alternate names, formulas, structures, physical data, medical and nonmedical uses. *The Merck Index* has a strong medical character but it is very useful for all users of chemicals.

WASTE DISPOSAL

The Resource Conservation and Recovery Act (RCRA) gives the U.S. Environmental Protection Agency (EPA) the responsibility to regulate chemical waste. In the laboratory, we must do our part to comply with all federal, state, and local regulations affecting disposal of wastes. Your laboratory instructor will discuss the procedures for waste collection. Wastes that cannot be discarded in the sink or wastebasket will be collected in special containers.

This manual minimizes the production of wastes by using smaller quantities of chemicals whenever possible.

SUGGESTED EQUIPMENT FOR EACH STUDENT LOCKER

Equipment	*Quantity*	*Check-In*	*Check-Out*
Apron	1		
Beaker, 150-mL	2		
Beaker, 250-mL	2		
Beaker, 400-mL	2		
Bottle, washing, plastic	1		
Brush, test-tube, large	1		
Brush, test-tube, small	1		
Cylinder, graduated, 10-mL	1		
Cylinder, graduated, 50-mL	1		
Crucible and lid, 15-mL	1		
Crucible tongs	1		
Dropper, medicine	4		
Evaporating dish, porcelain, 80-mm	2		
Flask, Erlenmeyer, 125-mL	2		
Flask, Erlenmeyer, 250-mL	2		
Forceps	1		
Funnel, 75-mm	1		
Reaction plate, 24-well	1		
Stirring rod, 200-mm	2		
Test-tube clamp	1		
Test-tube support (rack), wood or plastic	1		
Test tube, 10 × 75 mm	6		
Test tube, 16 × 150 mm	8		
Test tube, 25 × 150 mm	1		
Triangle	1		
Watch glass, 40-mm	1		
Watch-glass, 75-mm	2		
Wire gauze	1		

Nonreturnables

Filter paper, qualitative, 12.5-cm	1 pkg		
Litmus paper vial, blue	1		
Litmus paper vial, red	1		
Matches	1 box		

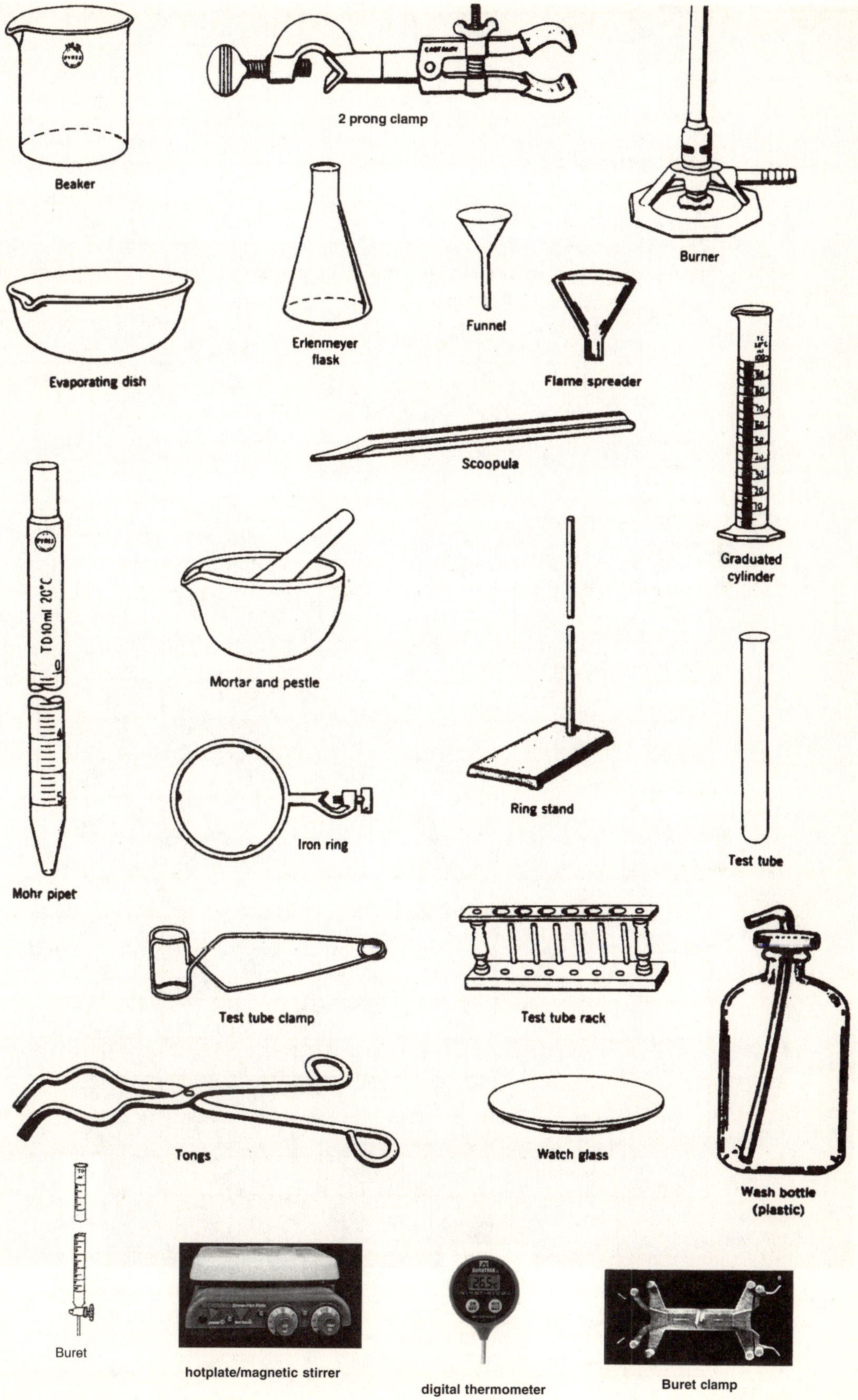
Beaker
2 prong clamp
Burner
Evaporating dish
Erlenmeyer flask
Funnel
Flame spreader
Scoopula
Graduated cylinder
TD 10ml 20°C
Mortar and pestle
Ring stand
Test tube
Iron ring
Mohr pipet
Test tube clamp
Test tube rack
Tongs
Watch glass
Wash bottle (plastic)
Buret
hotplate/magnetic stirrer
digital thermometer
Buret clamp

Fundamental Laboratory Operations

1. Handling Chemicals

Always read the label on a reagent bottle twice. Never remove the reagent bottle from the dispensing area. Never return excess chemicals to the reagent bottle. Never taste, touch, or smell chemicals unless specifically directed to do so. Hold a bottle with its label toward your palm to protect the label and the hand of the next user in case some reagent runs down the side of the bottle. Keep your hands and face clean. If a chemical inadvertently contacts your skin, wash the affected area with large amounts of water. Always wash your hands before leaving the laboratory.

a. Transferring Powders and Crystals

Tilt the jar slightly as illustrated in Figure 0-2, and roll the jar back and forth until the desired amount is obtained.

Withdraw only the quantity of chemical that you need, and never return any unused reagent to the bottle. To prevent contamination, lay the lid on the laboratory bench inner-side up. Do not insert a spatula into the bottle unless you are told to do so by your instructor. Replace the lid on the bottle even if others are waiting to use the bottle. If you spill any chemical, clean it up at once.

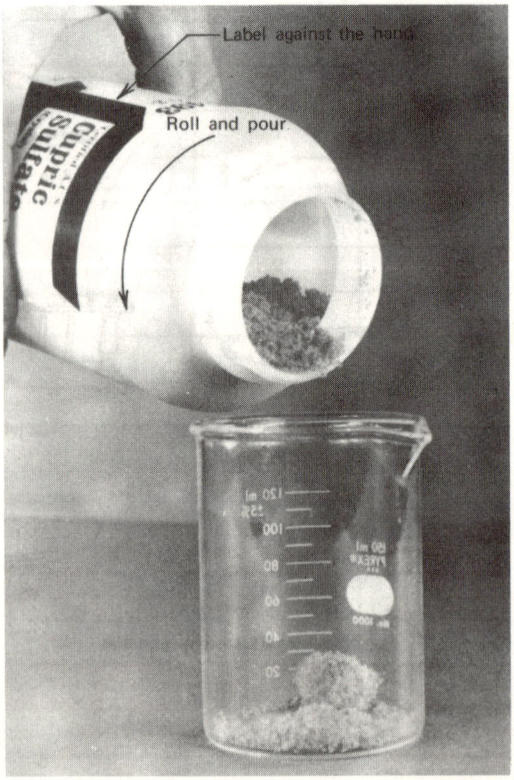

Figure 0-2 Dispensing a solid from a reagent bottle by rolling back and forth.

b. Transferring Liquids

If the bottle has a screw cap, simply unscrew the cap and lay it on the laboratory bench inner-side up. If the bottle has a ground glass stopper as in Figure 0-3, loosen the stopper by gently twisting it. Some ground glass stoppers have a vertical ridge which should be used to hold the stopper between two fingers as in Figure 0-4. Other ground glass stoppers have a flat top—these should lay on the laboratory bench inner-side up. Never allow the inner side of the stopper to touch the bench; impurities may be picked up which can contaminate the liquid when the stopper is returned. Replace the stopper on the bottle even if others are waiting to use the bottle.

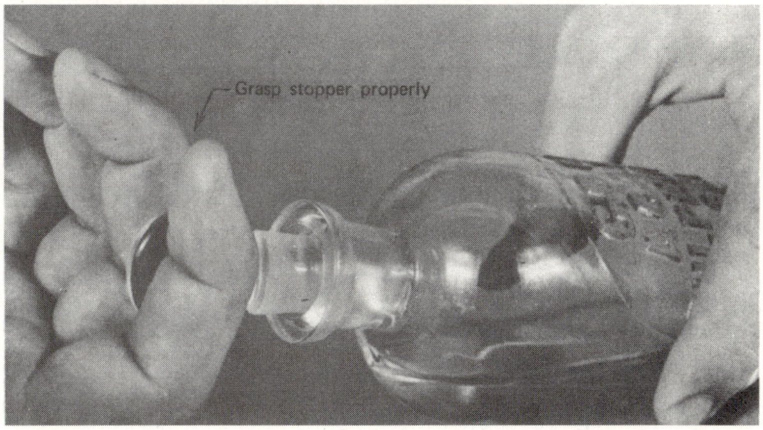

Figure 0-3 Remove the stopper by twisting and hold it between your fingers.

To transfer the liquid to another container, hold a stirring rod against the lip of the pouring bottle and pour down the stirring rod. The stirring rod should touch the wall of the receiving beaker as in Figure 0-4.

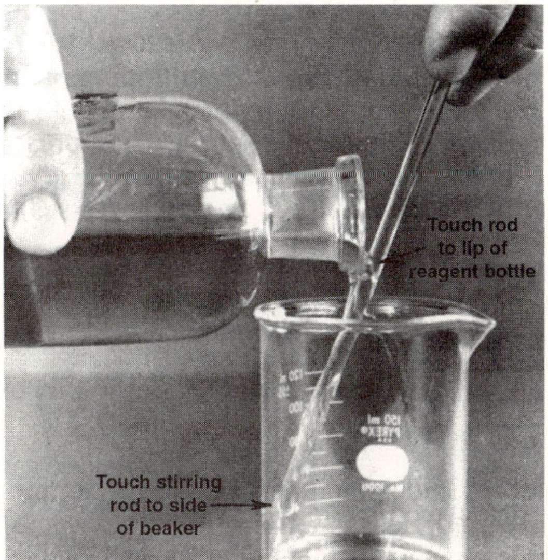

Figure 0-4 Use a stirring rod to transfer a liquid.

When transferring from a beaker, hold the stirring rod against the lip of the pouring beaker while touching the tip of the stirring rod to the side of the receiving beaker. See Figure 0-5.

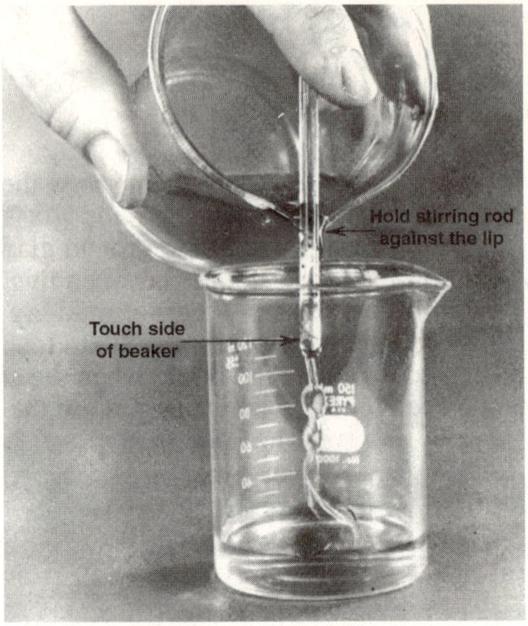

Figure 0-5 The stirring rod should touch the tip of the pouring beaker and the wall of the receiving beaker.

2. Cleaning Glassware

Glassware must be clean to prevent errors caused by contaminants. Wash the glassware with a small amount of detergent and tap water. Use a brush to scrub the glassware. Rinse the glassware several times with tap water, then rinse once or twice with a small amount of deionized* water. While rinsing with deionized water, tip and roll the glassware and allow the deionized water to wet the entire inner surface of the glassware. If rinsing a pipet, buret, or other glassware with a tip, discard the rinse water through the tip. If the glassware is truly clean, the final rinse water will flow smoothly down the glass without leaving any drops clinging to the glass.

Invert the clean glassware on a paper towel to dry. Use a test tube rack or other support to prevent test tubes or flasks from tipping over. If time does not permit air drying, a paper towel may be used to dry the glassware—this can cause contamination and ought to be avoided if possible. Do not use compressed air to dry glassware since it is not very effective and contamination is likely. Glassware can be placed in a drying oven to dry. Use insulated gloves in removing the glassware as it will be hot.

3. Use of Volumetric Glassware

Before use, all glassware should be thoroughly cleaned. See Fundamental Operation 2, above.

Graduated cylinders, pipets, burets, volumetric flasks, and syringes are the most commonly used glassware for measuring accurate volumes of liquids. Such glassware is available in an assortment of sizes, and each is used for a specific kind of laboratory operation. Beakers and flask are containers rather than measuring devices. The graduations on the sides of beakers and flasks provide only a rough estimate of volume and are not used for precise volume measurements.

---

*Deionized water is water that has been treated to remove ions (charged particles) and other impurities. Deionized water is very similar to distilled water.

All types of volumetric glassware have a slender, cylindrical shape in the measuring region, which causes the surface of most liquids to be curved downward. Take readings from the bottom of the curved surface (called the **meniscus**) with your eye at the same level. If the volumetric glassware that you are using has a number of graduations, estimate the volume as accurately as you can by noting the position of the meniscus between the graduations. See Figure 0-6.

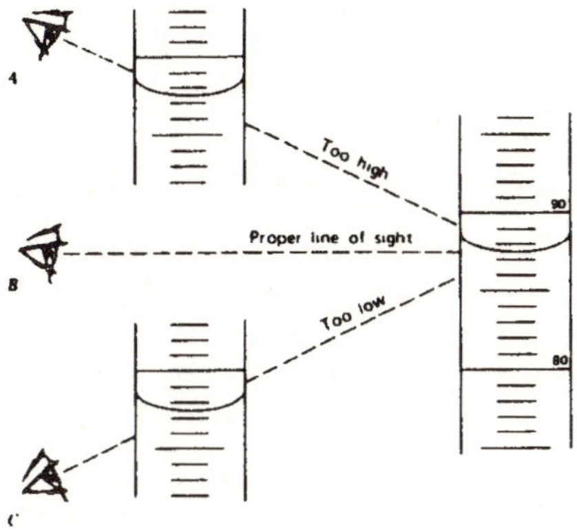

Figure 0-6 The proper method of reading the meniscus. The reading is 87.5 mL.

a. Graduated Cylinders

Graduated cylinders range in size from 1 mL to more than 1 L and provide a moderate degree of accuracy. They are the quickest and most commonly used of all volumetric glassware (see Figure 0-7). They are similar to kitchen measuring cups but their slender shape makes them more accurate. Before use, clean the cylinder as directed above.

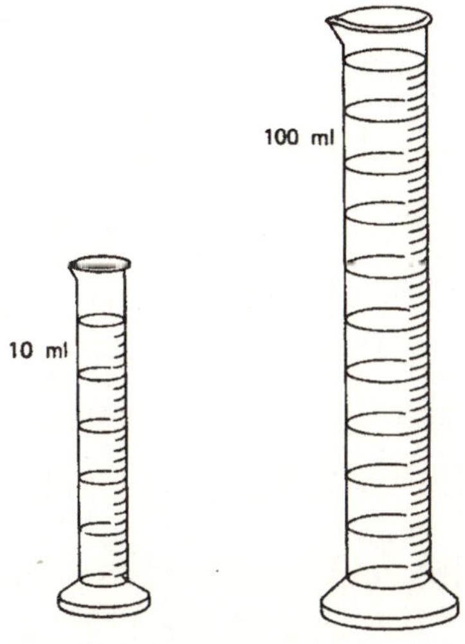

Figure 0-7 Graduated cylinders of different capacity.

b. Pipets

Pipets are useful for delivery of liquids when a greater degree of accuracy is required than is provided by a graduated cylinder. The two most frequently used pipets are the volumetric pipet and the serological pipet. See Figure 0-8. A volumetric pipet delivers only a single specific volume of liquid (for example 5 mL or 25 mL) and, therefore, has only one graduation mark. A serological pipet, on the other hand, has many graduation marks on its sides and may be used to deliver variable amounts of liquid.

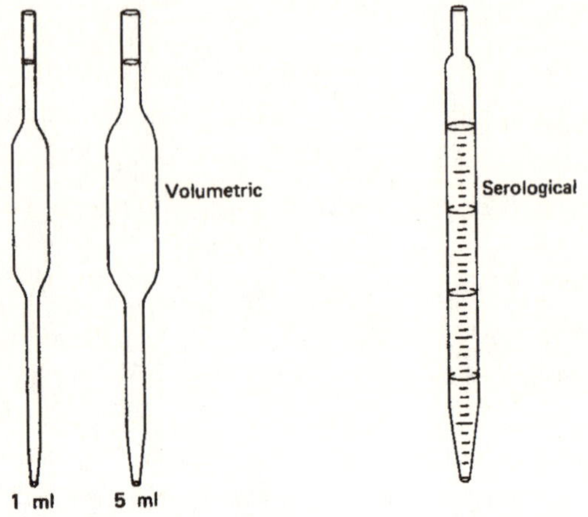

Figure 0-8 Volumetric and serological pipets.

Pipets marked "TC" (to contain) will contain the indicated volume when filled to the mark. Pipets marked "TD" (to deliver) will deliver the amount indicated on the pipets when the liquid flows out. A 10-mL TD pipet contains more than 10 mL—it delivers 10 mL and a small amount of liquid remains in the pipet. Volumetric pipets are usually marked "TD" and serological pipets are usually "TC."

In use, the liquid to be dispensed is drawn into the cleaned pipet by means of a pipet bulb; a variety of types are available. **Never use mouth suction to fill a pipet.** The liquid is drawn above the graduation mark, and the index finger is placed firmly on the opening at the top of the pipet. By carefully moving the index finger, the liquid is then allowed to drain slowly until the meniscus is exactly on the desired graduation mark. In the case of the volumetric pipet, the tip should be kept in contact with the side of the receiving container as the liquid drains into the container. A final drop is removed by touching the side of the container with the tip of the volumetric pipet. By design, a small amount of liquid will remain in the "TD" volumetric pipet; it is now "blown out." See Figure 0-9. With a "TC" serological pipet, when the entire contents is to be delivered, the last drop is "blown out" of the tip.

c. Burets

1. Description A buret is a hollow, narrow, cylindrical tube labeled with graduation marks and equipped with a regulating device (usually a stopcock) for controlling the flow of liquid from the tip. Since it has a narrow, uniform diameter, it is possible to deliver small quantities of liquid with a high degree of accuracy. The graduation marks on a buret begin with zero near the tip and increase down the buret. The final graduation mark is near, but not at, the bottom of the buret.

2. Preparation of the Buret Clean the buret with detergent and water before use. Do not pour solid detergent into the buret since it is apt to block the narrow tip; instead, dissolve a small amount of detergent in water and pour the solution into the buret. Use a long-handled buret brush to scrub the inside of the buret, but use extreme care to prevent the wire handle from scratching the inside walls of the buret. Rinse thoroughly with tap water, allowing the water to drain through the tip.

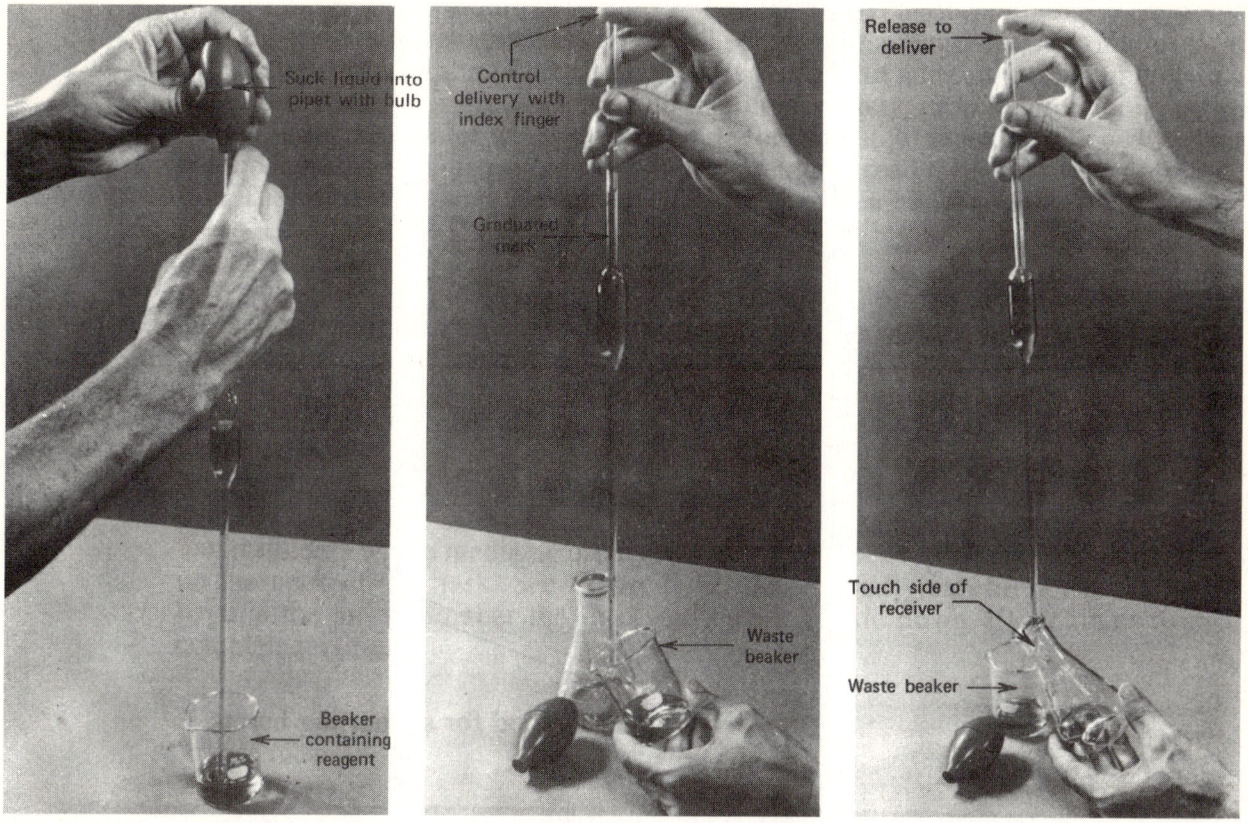

Figure 0-9 Delivering a liquid from a volumetric pipet.

Close the stopcock and fill with deionized water. Place a beaker or flask beneath the tip and allow the water to drain. Check to see that the water flows freely and that there are no leaks at the stopcock. Water droplets should not cling to the walls of the buret. Close the stopcock and add about 5 mL of the reagent solution (**titrant**). Tilt the buret and roll it so the titrant comes in contact with the entire inner surface. Place the buret in an upright position and allow the titrant to drain into a beaker or flask. Discard this titrant since it has been diluted by residual rinse water left in the buret.

Ask your instructor for assistance with a leaking stopcock.

Support the buret with a buret clamp. See Figure 0-10. Close the stopcock and fill to the top with the titrant to be used. Allow the titrant to drain through the tip until there is a solid column of liquid extending down to the tip, completely free of air bubbles.

3. Using the Buret The process of determining the amount of solution required to react with a given amount of sample is called a **titration**. The sample is placed in a receiving vessel (usually an Erlenmeyer flask). An **indicator**, a substance that changes color when the endpoint is reached, is added to the receiving flask. Place the flask on a sheet of white paper so that you can clearly see the color change of the endpoint.

Record the initial buret reading to the nearest 0.01 mL. It is not necessary for the initial reading to be exactly 0.00 mL—an initial reading of 2.45 mL or 3.10 mL is every bit as accurate.

Manipulate the stopcock with your left hand and swirl the flask with your right hand. (A left-handed person may find it more comfortable to switch hands.) See Figure 0-11.

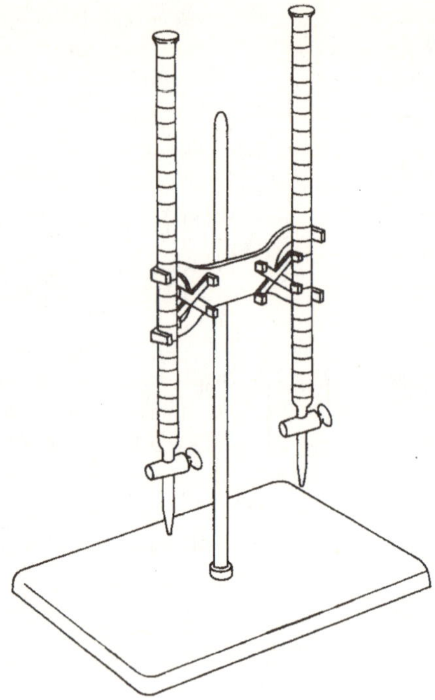

Figure 0-10 Proper method for supporting burets.

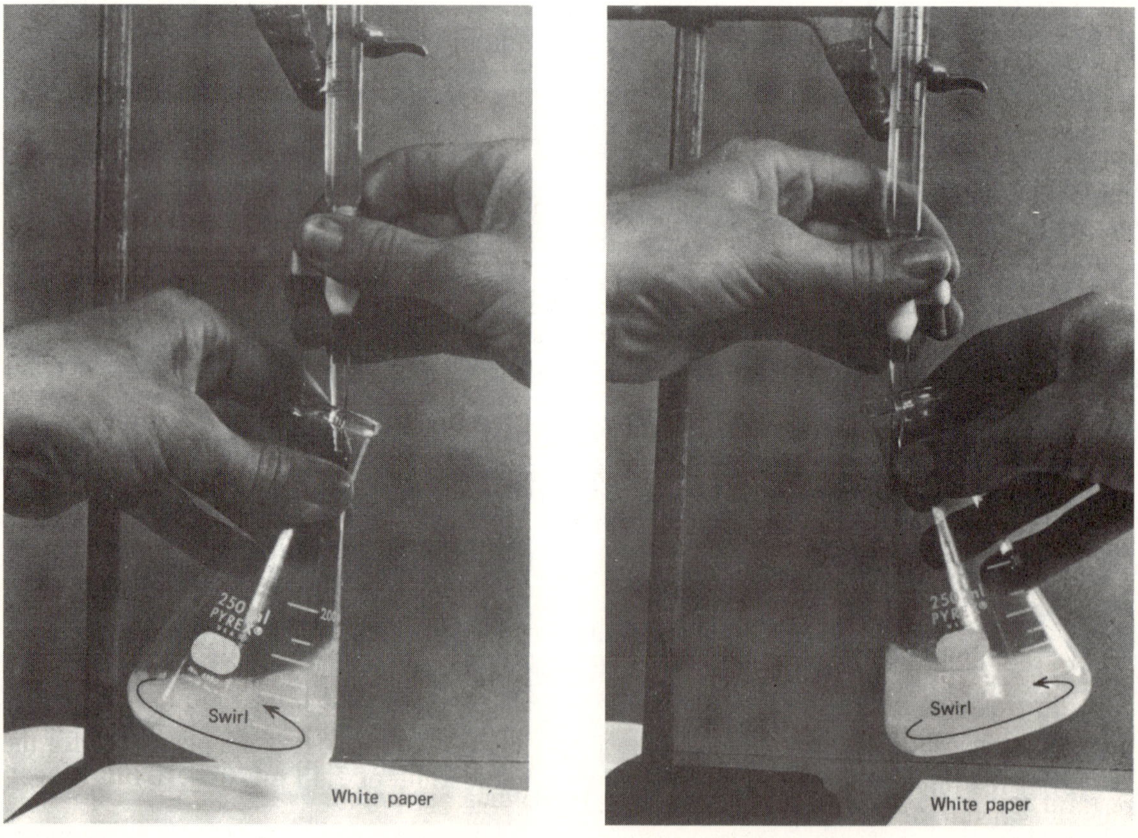

Figure 0-11 Correct titration procedure.

Dispense the titrant rapidly at first. You will notice that a color will appear momentarily at the spot in which the titrant comes in contact with the sample in the flask. As you near the endpoint, the indicator's color will persist over a longer interval of time. At this point, slow the addition of the titrant so that you do not add too much and go past the endpoint. While swirling the flask, add the titrant slowly, drop by drop until one drop finally produces a color which will not fade for at least 30 seconds. Allow 30 seconds for the titrant to drain from the walls of the buret; then read the buret to the nearest 0.01 mL. The volume of titrant used is the difference between initial and final readings.

d. Volumetric Flasks

These long-necked flasks are used to prepare solutions of specific concentrations. In practice, a measured amount of solute is placed in the clean flask. Add solvent until the flask is about half-full; it is much easier to swirl the solution and dissolve the solute when the flask is only half-full. Also, the volume of the solution often increases or decreases as the solute dissolves and, therefore, the solute must be totally dissolved before final filling of the flask to the graduation mark. Swirl the contents of the half-filled flask to completely dissolve the solute. Additional solvent is then added to the flask, the last few milliliters with a medicine dropper, until the meniscus reaches the graduation mark on the neck of the flask. See Figure 0-12. Be especially careful not to add solvent beyond the mark. Stopper the flask and turn it upside down 10-12 times to insure complete mixing.

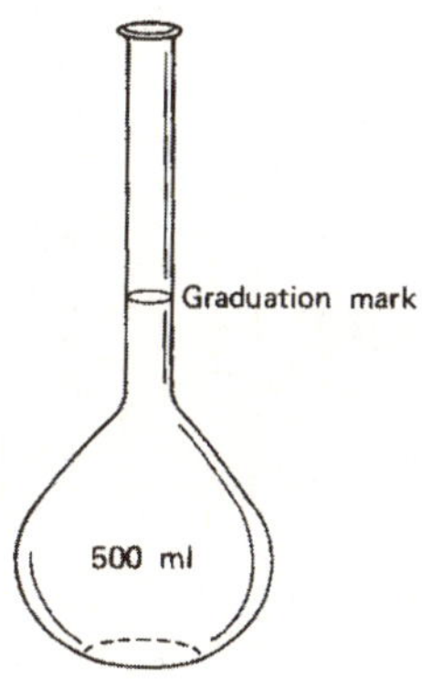

Figure 0-12 Volumetric flask.

4. The Laboratory Balance

When the mass of an object must be measured with an uncertainty of 0.01 g, a top-loading balance is frequently used. For higher precision (±0.001 g or ±0.0001 g) other electronic balances are available.

Some top-loading balances have a "tare" feature that is useful when finding the mass of a substance being added to a container. Enabling the tare function causes the balance display to read zero, even when a container is on the pan of the balance. When a substance is added to the container, the balance displays only the mass of the substance, not the combined mass of the substance and container.

You instructor will supply specific directions for the use of the balance in your laboratory. The following rules apply to every balance:

1. Never place a reagent directly on the weighing pan. Use either weighing paper or an appropriate container, such as a beaker.
2. Clean up all spilled materials.
3. Never return excess reagent to its original bottle. Your instructor will advise you as to its disposal.
4. If the balance appears to be out of adjustment, inform your instructor. Never attempt to make adjustments on your own.
5. Do not move the balance from one location to another. The balance must be readjusted after movement and movement may permanently damage the balance.

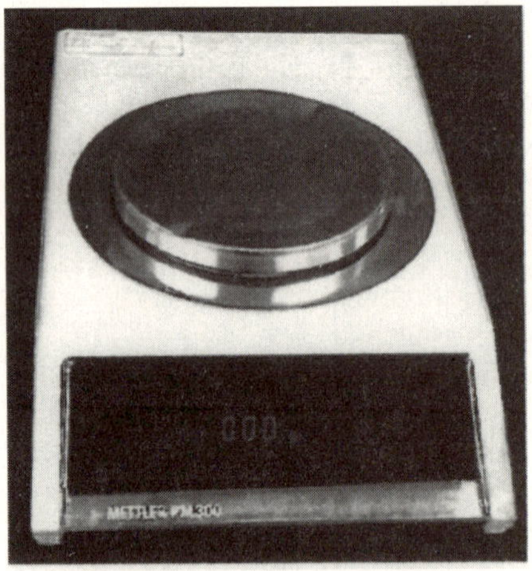

Figure 0-13 Top-loading balance.

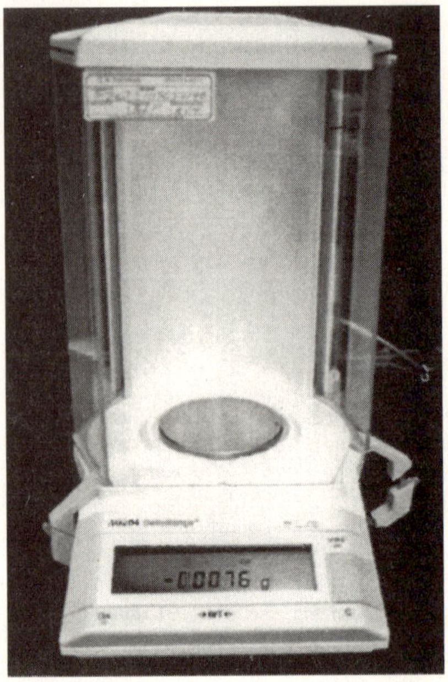

Figure 0-14 High-precision balance.

5. Laboratory Heat Sources

a. Bunsen Burner

1. Description

In 1855 the German chemist Robert Bunsen popularized a burner that mixed air with flammable gas to produce a hot, smokeless flame. Today, all laboratory burners are commonly referred to as Bunsen burners even if they are modifications of his design. The original Bunsen burner design did not have a gas adjustment valve. Most modifications, such as the Tirrill burner, have a gas adjustment screw at the base for controlling the supply of gas. The gas supply of the Bunsen burner must be controlled at the gas outlet valve on the laboratory bench—a less precise adjustment method. The supply of air entering the barrel of the burner is regulated either by adjusting a sleeve at the base or by screwing the barrel up or down on the base, depending upon which burner is used. See Figures 0-15 and 0-16.

The principal component of natural gas used in most laboratories is methane (CH_4). Because methane is odorless, a gaseous substance possessing a distinct odor is added to the methane so that gas leaks may easily be detected. When methane is burned in a sufficient supply of air, it combines with oxygen (O_2) to form carbon dioxide gas (CO_2) and water vapor (H_2O). The hot flame has two distinct blue cones and is nonluminous (does not produce light). See Figure 0-17.

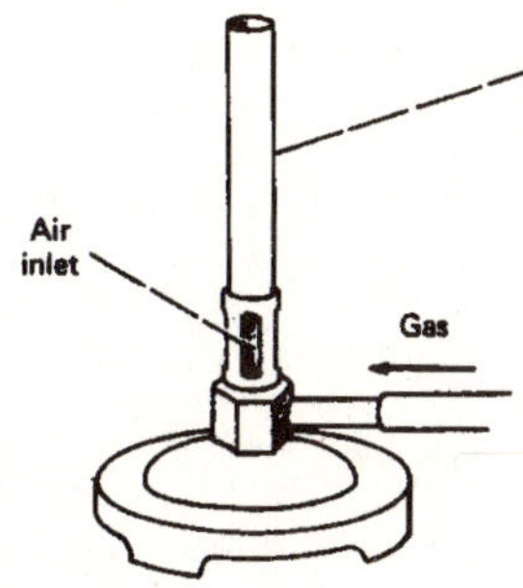

Figure 0-15 Bunsen burner.

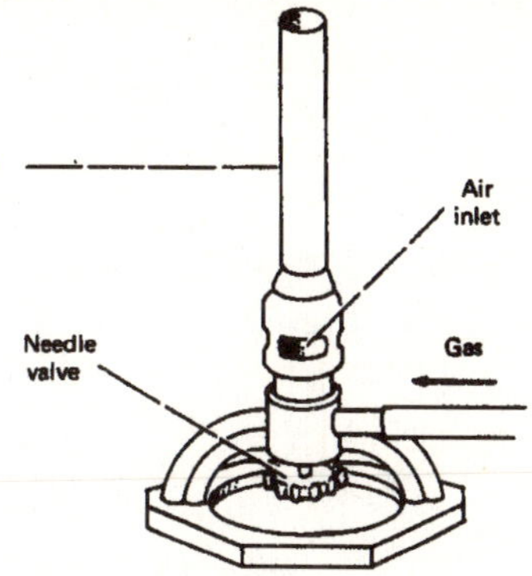

Figure 0-16 Tirrill burner.

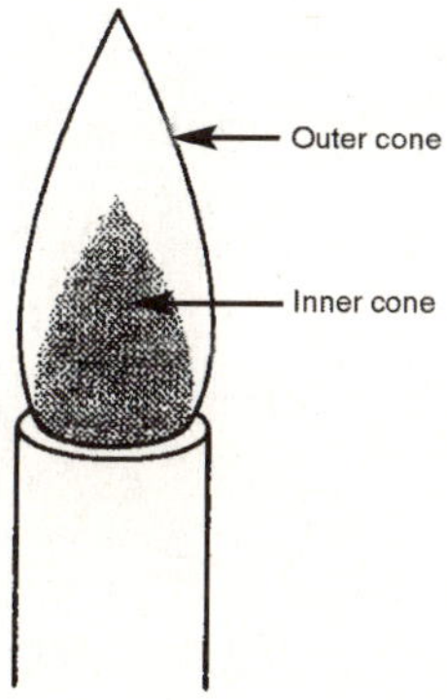

Figure 0-17 The flame of a properly adjusted Bunsen burner has two blue cones.

With an insufficient supply of air, the combustion of methane is incomplete—carbon (C) and carbon monoxide (CO) are produced. The tiny, hot, glowing pieces of solid carbon become incandescent and produce a yellow, luminous (light-producing) flame. Objects heated with a luminous flame become coated with black soot (unburned carbon). The carbon monoxide produced by incomplete burning of methane is a toxic, colorless, odorless gas.

2. Operation

Attach rubber tubing from the burner to the gas outlet on the lab bench. Fully close the air inlet on the burner, then open it slightly. Fully close the gas control valve on the burner, then open it slightly. Fully open the gas outlet on the lab bench. Bring a lighted match or friction lighter to the side of the burner at the top of the barrel and ignite the gas. Adjust the burner's gas control valve and air inlet for your experiment. For most laboratory operations, the flame should have two distinct blue cones and be about 8 cm (3 inches) tall. The flame should burn with a very slight "roar." Excessive "roaring" indicates the air inlet is open too far—this causes the gas to burn inside the barrel of the burner and the barrel will get hot.

b. Hotplates (Hotplate/Magnetic Stirrer)

1. Description

Electrical heat sources are another very common means used for heating in the laboratory. The most common type is probably a hotplate that may be equipped with a magnetic motor to provide even stirring during heating. One common example is shown in Figure 0-18. Those equipped with the stirring motor will have two controls. One dial controls the heating rate and the other dial the stirring rate. Most have a chemically resistant heating surface such as porcelain. Frequently the porcelain surface is white when cool and a yellowish color when hot so that visual observation will indicate the relative temperature of the surface. Because there is no flame associated with hotplates, they are somewhat easier and safer to use. These are often capable of providing temperatures up to 500°C if the heating is done without a water bath. Whether or not the container to be heated is placed directly on the heating surface, in a water bath, surrounded by an aluminum block or another method depends on what is being heated as well as the temperature needed.

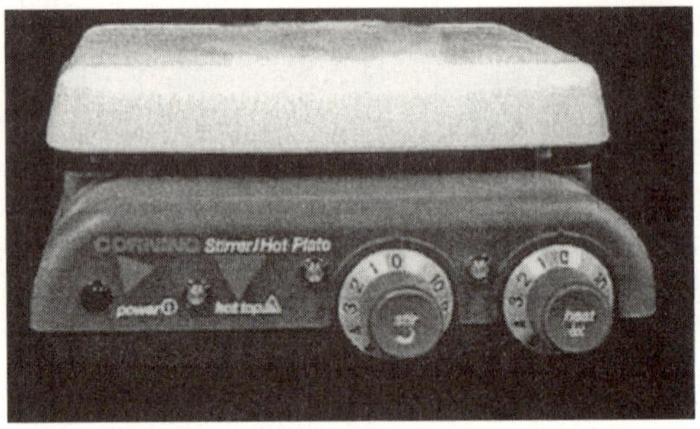

Figure 0-18 A laboratory hotplate/magnetic stirrer.

2. Operation

Place a flat wire gauze on the cool surface of the hotplate, then place the container or the bath that is to be heated on the center of the gauze. A magnetic stir bar or boiling stone should be placed in any liquid to be heated. Never add a boiling stone or stir bar to a hot liquid as this may cause extensive foaming and loss of the liquid out of the container. If the magnetic stir bar is used, turn the stirring motor on to provide even spinning of the stir bar. Boiling stones and the stir bar prevent hot spots from developing as the liquid is heated which can cause bumping of the liquid. Bumping can cause the liquid to be ejected from the container or cause the container to crack or tip over. When ready to start heating, the heat control is turned up slowly to obtain the desired temperature without overheating.

6. Heating Liquids

Whether a Bunsen burner or hotplate is used depends on the temperature needed. Bunsen burners can provide temperatures higher than those attainable with a hotplate.

a. In a Test Tube

1. Using a Bunsen Burner

The test tube should not be more than one-third full. Clamp the tube at the top so as to provide a large heating surface. Incline the test tube slightly, and direct the flame at the top of the liquid, not at the bottom. Move the test tube constantly. See Figure 0-19.

CAUTION: Do not point the open end of a test tube toward another person—the liquid may suddenly shoot out if the test tube is not properly heated. Do not heat a closed or stoppered container—pressure will build up in the container. Never heat graduated cylinders, bottles, watch glasses, or other thick-walled glassware—they will break.

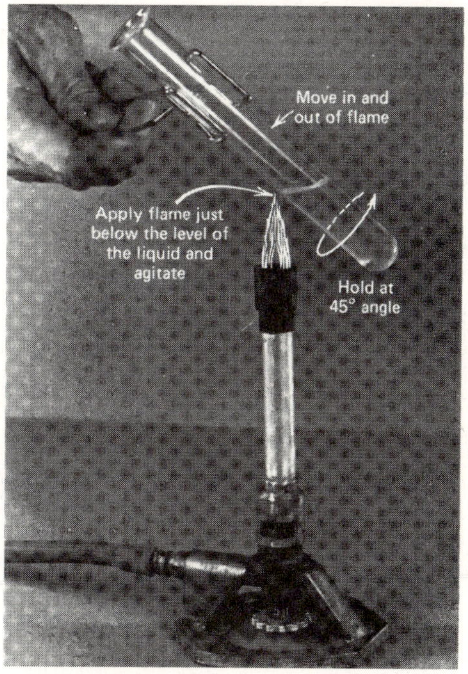

Figure 0-19 Heating a liquid in a test tube.

2. Using a Hotplate

Test tubes are usually heated in a bath (commonly water). The test tube is supported in the bath to prevent tipping. A simple apparatus for doing this is shown in Figure 0-20. The beaker is partially filled with water, a boiling stone or stir bar is added, and a piece of aluminum foil is placed over the top of the beaker and a small hole is made in the foil with the end of a spatula. The test tube is fitted through the hole and is supported by the foil. The stirring motor, if applicable, is turned on to obtain even stirring and then the heat control is adjusted to obtain the desired temperature.

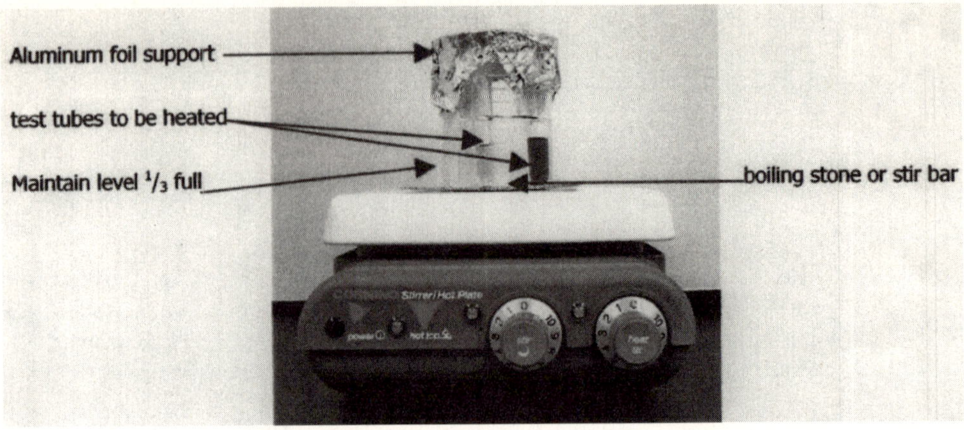

Figure 0-20 Water bath containing test tubes.

b. In a Beaker or Flask

1. Using a Bunsen Burner

Set up a ring-stand with iron ring and wire gauze as illustrated in Figure 0-21. A second iron ring prevents the beaker from falling off the wire gauze. Place a stirring rod in the liquid to avoid violent bumping or spattering of the liquid. Position the flame beneath the tip of the stirring rod.

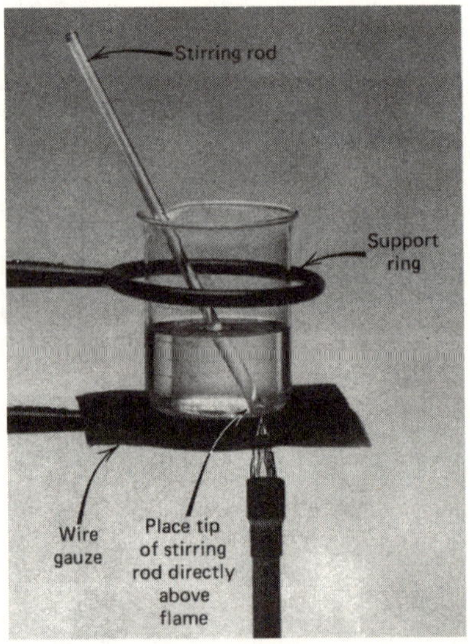

Figure 0-21 Proper method for heating a liquid in a beaker.

2. Using a Hotplate

Add a magnetic stir bar or boiling stone to the liquid to be heated in a beaker or flask. Usually a watch glass cover is placed on the container to prevent evaporative loss of the liquid while not giving an airtight seal. Place the container on the center of the wire gauze on the hotplate. Start the stirrer motor, if present. Again slowly turn the heating control to obtain the desired temperature. This set up is shown in Figure 0-22.

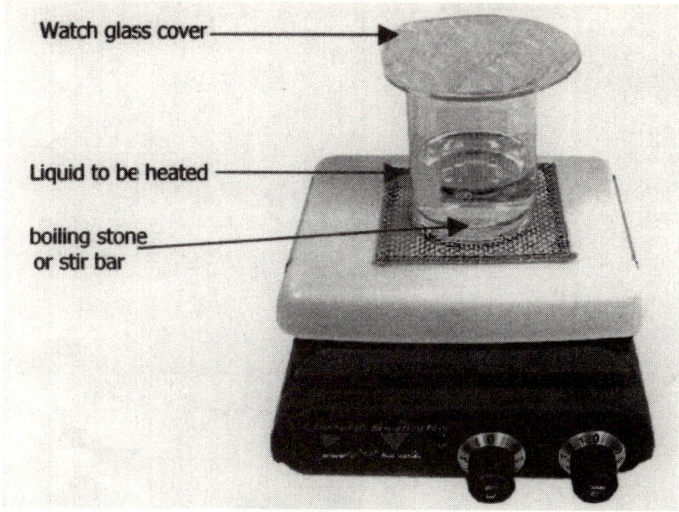

Figure 0-22 Heating a liquid in a beaker.

7. Evaporating Liquids

CAUTION: Do not use a flame near flammable liquids. Use a fume hood if noxious odors are likely to be produced.

1. Using a Bunsen Burner

a. Over an Open Flame

Nonflammable liquids may be evaporated from an evaporating dish using a gentle, "cool" flame as illustrated in Figure 0-23. When the quantity of liquid remaining in the evaporating dish becomes very small, remove the burner to avoid spattering.

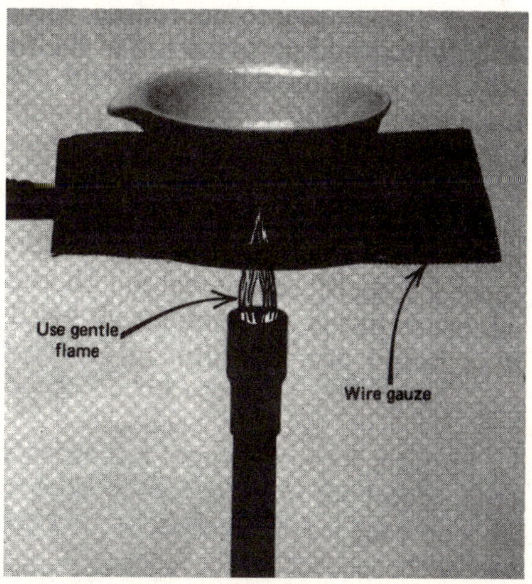

Figure 0-23 Evaporation of a nonflammable liquid over a direct flame.

b. Over a Steam Bath

Nonflammable liquids may also be evaporated from a steam bath. The evaporating dish is placed over a beaker of boiling water. See Figure 0-24. The steam bath provides a moderate, uniform distribution of heat that prevents burning, scorching, or overheating.

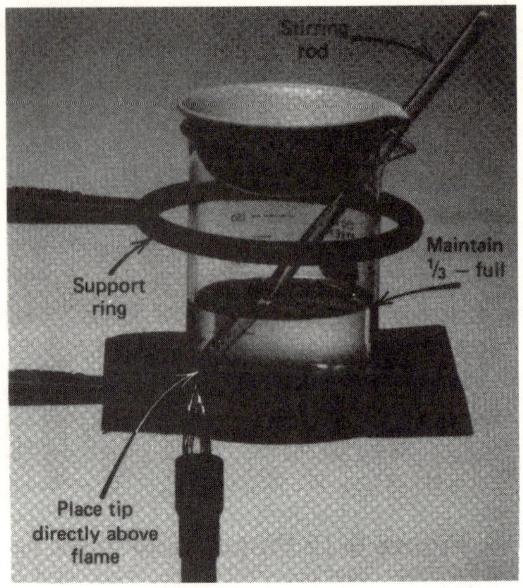

Figure 0-24 Evaporation of a nonflammable liquid over a steam bath.

b. Using a Hotplate

Usually liquids are evaporated using a water bath on a hotplate. The beaker that will serve as the bath is partially filled with water and a boiling stone or magnetic stir bar is added. The container, usually an evaporating dish, with the liquid to be evaporated is placed on top of the beaker. The spout of the beaker should allow air to enter or water vapor to leave the beaker so that an airtight system is not set up. See Figure 0-25. The stirring motor, if present, is turned on for even stirring and the heat control turned up to allow the water to boil. The water vapor carries the heat to the bottom of the container causing the liquid to evaporate. Do not allow the water bath to go to dryness nor the container being heated. When the liquid has nearly evaporated from the container, with tongs, remove the container with the sample and place it on another wire gauze to finish evaporating and to subsequently cool on the bench top.

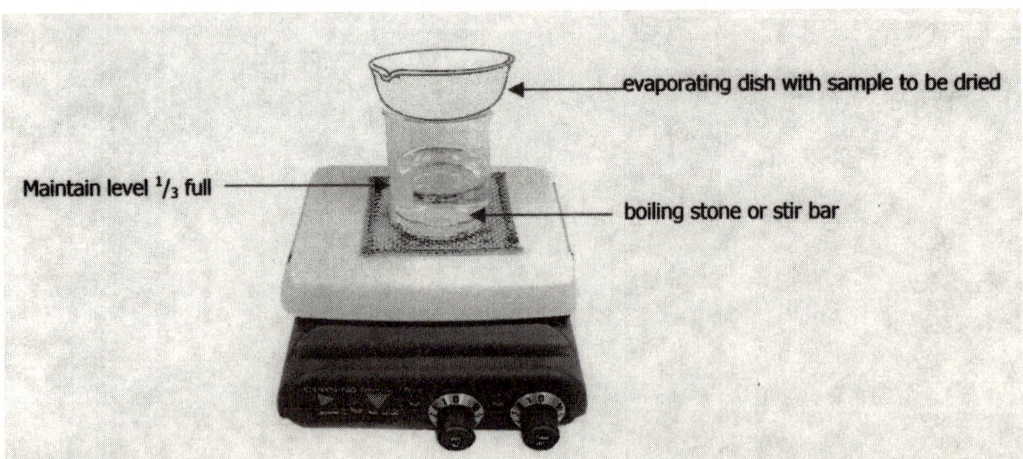

Figure 0-25 Evaporating a liquid.

Solids isolated by filtration can be dried in a similar manner by opening the filter paper and placing it on a watch glass. The watch glass can then be placed on top of the water bath. See Figure 0-26.

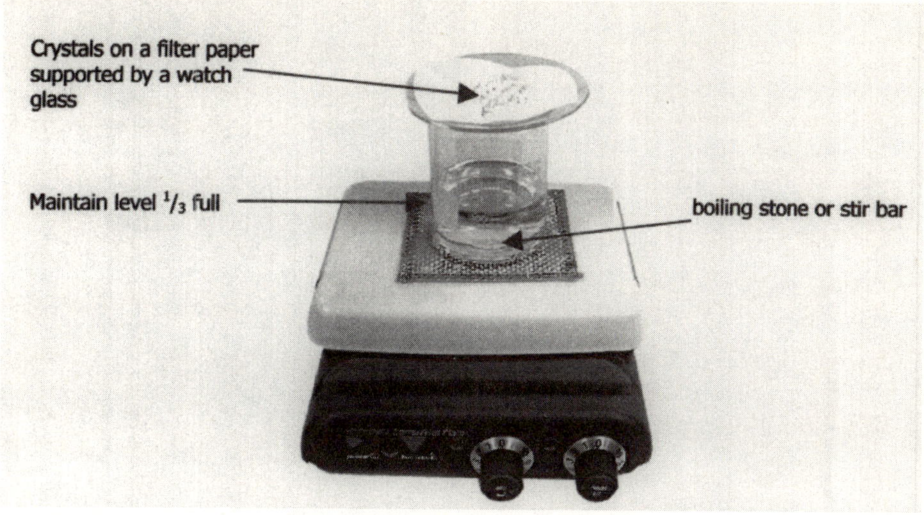

Figure 0-26 Crystals on filter being dried.

8. Filtration

Figure 0-27 illustrates how to fold a filter paper and arrange the paper for filtration. The paper is first folded in half. Then fold the paper again, not quite in perfect quarters. The two folded edges should not quite touch; the second edge should be about 3 mm (one-eighth inch) from the first edge. Do not crease the folds too sharply or a hole may develop at the point of the paper. A small corner is torn from the smaller section. The tear enables the paper to make a tight seal against the funnel. The paper is then opened to form a cone and inserted into the funnel. Moisten the paper with a small amount of solvent (usually water) and press the top of the filter paper against the funnel to make a tight seal.

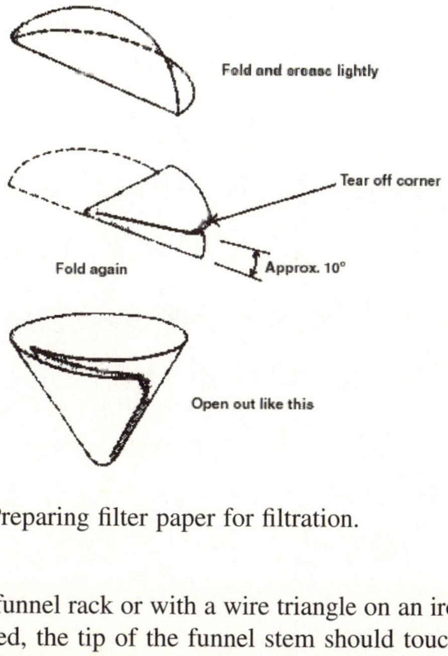

Figure 0-27 Preparing filter paper for filtration.

The funnel should be supported with a funnel rack or with a wire triangle on an iron ring. To avoid splashing and to increase filtration speed, the tip of the funnel stem should touch the side of the receiving vessel. See Figure 0-28. Never fill the funnel more than two-thirds full of solution.

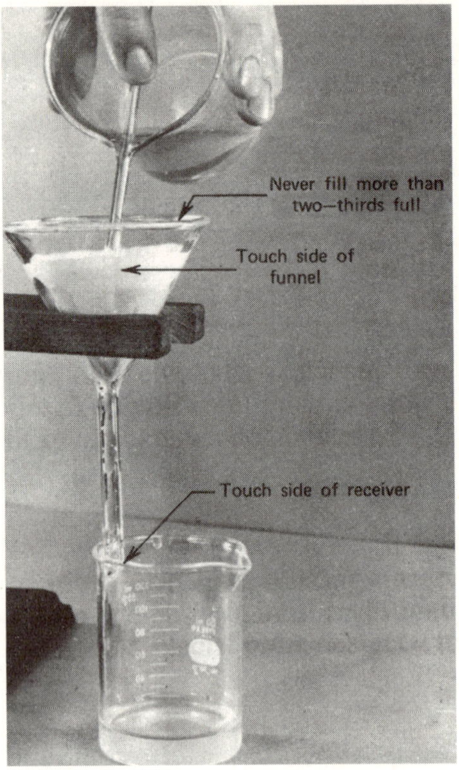

Figure 0-28 Proper filtration set-up.

9. The Laboratory Thermometer

Laboratory thermometers can be of the liquid-filled type. For routine use, most laboratories use spirit-filled thermometers, filled with hexane, mineral oil, or alcohol, which is dyed red for visibility.

Laboratory liquid-filled thermometers do not have to be "shaken down" before use. In fact, shaking or dropping a liquid-filled thermometer is apt to cause the liquid column to separate, meaning there is a gap in the liquid. If the liquid column is separated, give the thermometer to your instructor for repair.

Digital thermometers are replacing liquid-filled thermometers. In general, they are more precise and many models are quite inexpensive. The dials of these thermometers usually display the temperature of the sample to ±0.1 or 0.2°C. They have a fairly rapid equilibration time and are not subject to the column separation problems and breakage of liquid-filled thermometers. They also cover a wider temperature range than that of many liquid-filled thermometers. They do require a battery. Operation is straightforward and usually only requires pushing a button on the face of the thermometer. An example of a digital thermometer is shown in Figure 0-29.

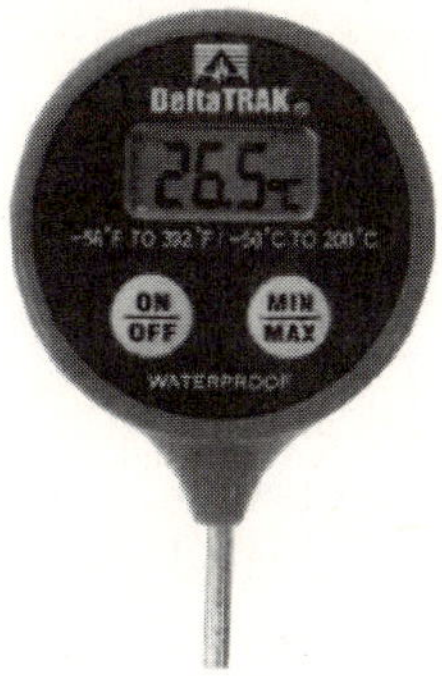

Figure 0-29 A simple inexpensive digital thermometer.

You might need to measure the temperature of a liquid while you are heating it. The thermometer should be held in the center of the liquid. Do not rest the thermometer on the bottom of the beaker since the glass beaker may be a different temperature than the liquid it contains.

A thermometer must be allowed time to adjust to the temperature of the substance being measured. A laboratory thermometer usually responds rapidly, perhaps within 30 seconds depending on how much the substance's temperature differs from the thermometer's initial temperature. Watch the liquid column move; after it stops moving wait at least another 30 seconds before taking a temperature reading. For digital thermometers, monitor the display until it reaches a steady value.

Experiment 1
Laboratory Techniques

Purpose: To familiarize you with the Bunsen burner and the processes of filtration and evaporation.

Introduction

LABORATORY BURNERS

So that you become familiar with the operation of the Bunsen burner, carefully read Part 5 of Fundamental Laboratory Operations, page 16.

FILTRATION

The process usually employed to separate an insoluble solid from a liquid is called **filtration**. Industrially the filtration process is of great importance. Many large cities use layers of rock beds to filter water for drinking. Some large breweries use millipore filters made of cellulose acetate in preparing draft beer for bottling. These millipore filters are so fine that harmful bacteria are filtered out and flavor-altering heat-pasteurization is not required.

In the laboratory the filtration process is carried out using a funnel and filter paper. Refer to Part 8 of the Fundamental Laboratory Operations on page 23. The liquid that passes through the filter paper is called the filtrate, and the solid that remains on the filter paper is the **precipitate** or **residue**.

EVAPORATION

Most dissolved substances (**solutes**) can be recovered from solution by evaporating the liquid portion (**solvent**) over a direct flame, steam bath, or on a hotplate. Figures 0 23 and 0 24, pages 21 and 22, illustrate typical evaporation setups.

THE CHEMICAL REACTION

Many chemical reactions taking place in **aqueous** (water) solutions frequently form insoluble solids as products. These precipitates may be separated from water-soluble solids through filtration. In Part II of this experiment, aqueous solutions of two reactants—strontium chloride and sodium sulfate—react to form two products: strontium sulfate and sodium chloride.

strontium chloride + sodium sulfate → strontium sulfate + sodium chloride

(reactants) (products)

Both of the reactants and one of the products are water-soluble. The other product (the precipitate) is insoluble in water and may be separated by filtration. The water-soluble product may be isolated by evaporation of the water.

Strontium chloride is used in pyrotechnics (fireworks) to produce red color. It is also used as a dental desensitizer under the name *Elecol*. Sodium sulfate is sometimes known as Glauber's salt and is used medically as a cathartic to purge the bowels. Other uses include dyeing and printing fabrics and the manufacture of glass.

Procedure

Use a pen or pencil to check off each step as they are completed.

CAUTION: Make sure you wear safety goggles. Also, make sure all loose clothing and long hair is secured/tied back before lighting the burner.

1. LABORATORY BURNERS

1. Examine your burner. Make sure you can identify the parts labeled in Figures 0-15 and 0-16, page 17.

2. Attach rubber tubing from the burner to the natural gas outlet on the lab bench. Fully close the air inlet on the burner, then open it slightly. Fully close the gas control valve on the burner, then open it slightly. (Remember that the burner is turned off using the gas valve on the bench top. Do not use the gas control on the burner.)

3. Fully open the gas outlet on the lab bench. Bring a lighted match or friction lighter to the side of the burner at the top of the barrel and ignite the gas.

4. Adjust the burner's gas control valve and air inlet. The normal, proper flame should have two distinct blue cones and be about 8 cm (3 inches) tall. The flame should burn with a very slight "roar." Excessive "roaring" indicates the air inlet is open too far—this causes the gas to burn inside the barrel of the burner, and the barrel will get hot.

5. Use a white piece of paper as a background to help you see the flame better and sketch the flame on the Report Sheet. Use caution to avoid catching the paper on fire.

6. Completely close the air inlet. This flame is not normal or proper. Sketch and indicate the color of the flame on the Report Sheet.

7. Without adjusting the flame (air inlet closed), use your crucible tongs to hold an evaporating dish in the top part of the flame for about 1 minute. See Figure 1-1. Describe the appearance of the underside of the evaporating dish on the Report Sheet.

8. Allow the evaporating dish to cool. Clean the dish by first wiping with a dry paper towel—then wash with a brush and detergent, rinse and dry.

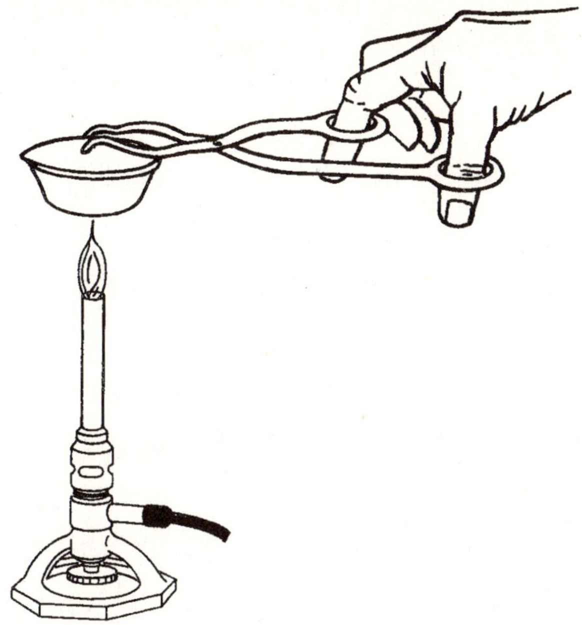

Figure 1-1 Evaporating dish held over a burner flame.

9. Adjust the burner by opening the air inlet to produce a normal, proper flame. The flame should be about 8 cm (3 inches) tall and should have a sharply defined inner blue cone.

10. Use your crucible tongs to hold an evaporating dish in the top part of the flame for about 1 minute. Describe the appearance of the underside of the evaporating dish (with air inlet open) on the Report Sheet.

11. Hold a piece of iron wire with you forceps and place the wire in region *a* (top of outer blue cone) as indicated in Figure 1-2 for about 30 seconds. Observe how brightly the wire glows.

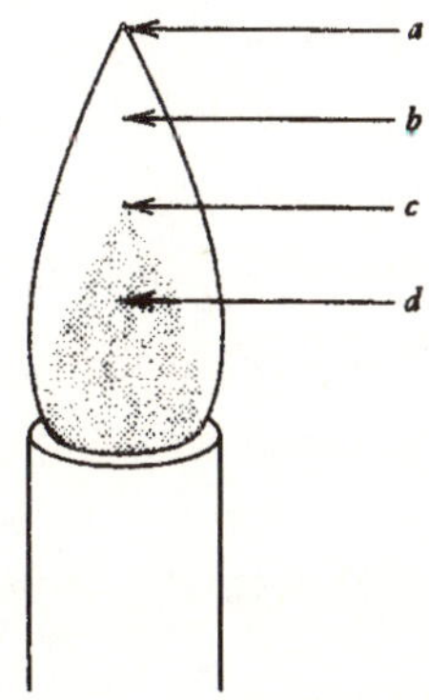

Figure 1-2 Regions of flame for temperature measurement.

12. Move the wire to region *b* (between the top of the inner cone and the top of the outer cone) for about 30 seconds. Then move to region *c* (top of inner cone) and region *d* (within the inner cone) for 30 seconds each. Observe how brightly the wire glows in each position. Record your observations on the Report Sheet. Allow the wire to cool and dispose of as directed by your instructor.

WE RECOMMEND THAT STEPS 13-17 OF THIS EXPERIMENT ARE CONDUCTED BY AN INSTRUCTOR FOR DEMONSTRATION PURPOSES ONLY, DUE TO THE FIRE DANGER INVOLVED WITH THESE PROCEDURES.

13. Turn on the water at the tap. Use your forceps or crucible tongs to quickly insert a piece of cardboard vertically into the flame. The cardboard should rest on the top of the barrel of the burner and split the flame—half of the flame should be on each side of the cardboard. See Figure 1-3. Hold the cardboard in this position until it becomes heavily blackened but do not allow the cardboard to burst into flames. Quickly cool the cardboard in running water. Sketch the blackened pattern on the Report Sheet.

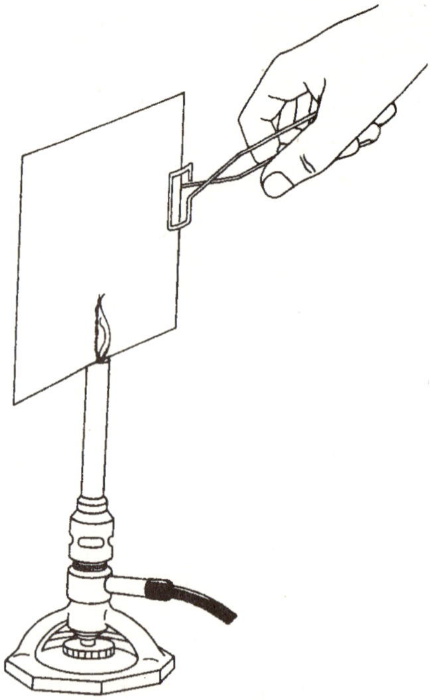

Figure 1-3 Cardboard held vertically in burner flame.

14. Without adjusting the flame, turn off the burner by closing the gas outlet valve on the bench top. Allow the burner to cool.

15. Obtain either a bobby pin or a straight pin. If using a bobby pin, push it half-way onto the match immediately below the head of the match. See Figure 1-4. If using a straight pin, insert the pin through the stem of the match immediately below the head of the match. See Figure 1-5.

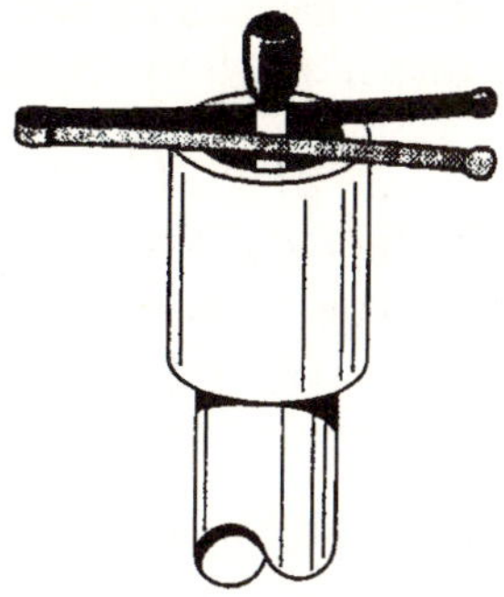

Figure 1-4 Match supported by bobby pin in barrel of burner.

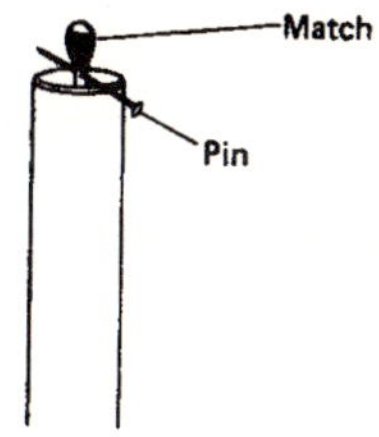

Figure 1-5 Match supported by straight pin in barrel of burner.

16. Hang the bottom of the match in the center of the barrel of the burner, supported by the bobby pin or straight pin. Make sure the match is centered and the match head is immediately above the top of the barrel of the burner. Fully open the gas outlet valve on the bench top and light the burner. Describe the behavior of the match and explain the results.

17. Turn off the burner by closing the gas outlet valve on the bench top. Never turn off a burner with the gas control valve on the burner.

II. FILTRATION

1. Review Part 1b of Fundamental Laboratory Operations on pages 9-10 for proper transferring of liquids. Fill one 16- × 125-mm test tube (the middle-sized test tube) half-full (about 10 mL) of strontium chloride solution (0.2 *M**) and a second test tube half-full (about 10 mL) of sodium sulfate solution (0.2 *M*).

2. Pour both solutions into a 150-mL beaker and stir with a glass stirring rod. Wait 2 minutes before filtering to allow the precipitate to form and coagulate.

3. Prepare a filtration step as shown in Figures 0-27 and 0-28, pages 23 and 24. Fold a filter paper in half, then fold again not quite at a right angle. Tear off a tiny piece of the corner of the smaller layer of paper. Open the paper into a cone and place the paper snugly into the funnel. Moisten the paper with a few milliliters of deionized water and press the paper against the top of the funnel to form a tight seal. Support the funnel with a funnel rack, iron ring, or clamp. The tip of the funnel should touch the side of a 100-mL receiver beaker.

4. Filter the solution produced in Step 2. Never fill the funnel more than two-thirds full of solution. The filtrate should be clear; if it is not, filter it again. After the liquid has drained through the filter paper, use a small stream of deionized water from the wash bottle to transfer any remaining precipitate from the beaker to the filter paper. Do not use more than a few milliliters of wash water since you will later evaporate the water.

5. Save the clear filtrate for use in Part III. Describe the color and appearance of the precipitate on the Report Sheet. Discard the filter paper in the container labeled "Waste Precipitate" or as directed by your instructor.

---

*0.2 *M* describes the concentration of the solution.

6. The reaction of strontium chloride and sodium sulfate formed two products; one product is soluble and the other is insoluble in water. To identify the filtrate and the precipitate, test the solubilities of the two products—strontium sulfate and sodium chloride. Review Part 1a of Fundamental Laboratory Operations on page 8 for proper transferring of powders and crystals. Place a small amount (only a few crystals) of sodium chloride into a regular (16- × 125-mm) test tube half-filled with deionized water. Stopper the test tube and thoroughly mix the contents. Observe the solubility of sodium chloride in water. Discard the sodium chloride in the sink.

7. Place a small amount (only a few crystals) of strontium sulfate into a 16 × 125-mm test tube half-filled with deionized water. Stopper the test tube and thoroughly mix the contents. Observe the solubility of strontium sulfate in water. Discard the strontium sulfate in the "Waste Salts" container or as directed by your instructor.

8. Identify the filtrate and precipitate based on the solubility tests.

III. EVAPORATION

1. Adjust the ring on the ring stand so that it is about 8 cm (3 inches) above the top of the barrel of the burner. Place a wire gauze on the ring. To use a hotplate for evaporation see Figure 0-25, page 22.

2. Pour about half the filtrate obtained in Part II into an evaporating dish and evaporate to dryness over an open flame (see Figure 0-23, page 21), or over a water bath on a hotplate (see Figure 0-25, page 22). Allow the solution to slowly boil. As the evaporating dish nears dryness, reduce the heat to prevent spattering. **Make sure you wear safety goggles.**

3. Describe the color and appearance of the residue. The size of crystals formed depends on how rapidly they were formed. Formed rapidly, crystals are very tiny and are often called a powder. Formed more slowly, the crystals are larger.

4. Allow the evaporating dish to cool in the air. Cold water will crack hot porcelain.

IV. CLEANUP

1. If possible, allow the wire gauze, iron ring, and ring stand to cool in the air. If time does not allow air-cooling, use running water to cool them before putting them away. For a hotplate, allow the wire gauze and the heating surface to cool before putting them away.

2. Discard the remaining filtrate in the sink and rinse the evaporating dish residue into the sink.

3. Wash the glassware with a brush and detergent. Rinse well with tap water. Rinse once or twice with a small amount of deionized water. Drain and allow to air dry.

4. Wash your hands before leaving the laboratory.

Name ______________________ Section ____________ Date ____________

Prelab Questions for Experiment 1

_______ 1. The natural gas in most laboratory burners is primarily
a. oxygen
b. carbon dioxide
c. butane
d. methane

_______ 2. To light a laboratory burner, hold a match
a. about 3 inches above the top of barrel
b. about 1 inch above the top of barrel
c. to the side of the burner at the top of the barrel

_______ 3. A yellow, luminous flame indicates the laboratory burner
a. needs additional air
b. needs less air
c. is properly adjusted

_______ 4. A flame with two distinct blue cones and loud "roaring" indicates the laboratory burner
a. needs additional air
b. needs less air
c. is properly adjusted

_______ 5. To turn off a laboratory burner,
a. close the gas outlet valve on the bench top
b. close the gas control valve on the burner
c. close the air inlet on the burner

_______ 6. The best way to separate sugar from water is
a. filtration
b. evaporation

_______ 7. The best way to separate sand from water is
a. filtration
b. evaporation

_______ 8. In a solution of salt in water,
a. the solute is water and the solvent is salt
b. the solute is salt and the solvent is water

9. In Part II of this experiment, which are the reactants?

______________________ and ______________________

and which are the products?

______________________ and ______________________

10. Define *filtrate*.

11. Define *precipitate*.

Name ______________________ Section ____________ Date ______________

Report for Experiment 1

I. LABORATORY BURNERS

1. Sketches of flames with color observations:

   a. Air inlet open

   b. Air inlet closed

2. Appearance of underside of evaporating dish:

   a. Air inlet open ______________________

   b. Air inlet closed ______________________

3. Why did the underside of the evaporating dish appear different with the air inlet closed? Hint: See Fundamental Laboratory Operations, Part 5a.1, page 16.

   ____

4. In which region of the flame (*a, b, c*, or *d*) did the iron wire glow most brightly?

   ____

5. In which region of the flame (*a, b, c*, or *d*) did the iron wire glow least brightly?

   ____

6. Sketch the charring pattern on the cardboard and label the area where no charring took place.

7. Describe the behavior of the match suspended on the barrel of the burner.

____

____

8. Is the behavior of the match consistent with your observations of the glowing iron wire (Part I, Steps 11 and 12) and the charring of the cardboard (Part I, Step 13)?

____

____

9. Explain why the match behaved as it did.

____

____

10. If you wished to quickly heat a beaker of water on a ring stand, how would you adjust the height of the iron with respect to the flame of the Bunsen burner?

____

____

Name ______________________________ Section ____________ Date ______________

II. FILTRATION

1. Solubility Tests:

    a. Solubility of sodium chloride in water (soluble or insoluble?) ________________

    b. Solubility of strontium sulfate in water ______________________________

2. What substance is contained in the filtrate? ______________________________

3. What substance is the precipitate? ______________________________

4. How were you able to identify the filtrate and the precipitate? ________________

____

III. EVAPORATION

1. Describe the color and appearance of the residue left in the evaporating dish.

____

____

Related Questions for Experiment 1

1. LD_{50} is a measure of toxicity—lethal dose, 50 percent kill. LD_{50} is the dose in mg of the substance which is most likely to cause death within 14 days in one half of the animals used in the testing per kilogram of body weight. The LD_{50} of barium chloride is 118 mg/kg and the LD_{50} of strontium chloride is estimated at 2060 mg/kg. Which compound is more toxic?

____

2. Barium compounds are chemically very similar to strontium compounds. Water-soluble compounds of both are toxic, but to different degrees. An antidote for barium chloride poisoning is to orally administer 30 grams of sodium sulfate in 250 mL of water. Explain the chemical basis for this antidote.

____

____

Experiment 2
Measurement

Reference: *General, Organic, and Biological Chemistry: An Integrated Approach, Chapter 1.*

Purpose: To familiarize you with the measurement process and use of the balance, graduated cylinder, thermometer, and urinometer.

Introduction

MEASUREMENT SYSTEMS

Nearly all the world's people use a set of measurement units called *Le Système International de d'Unités*, commonly referred to as the SI system. The SI system is a recent modification of the older metric system. The SI system is popular because it is a decimal number system. Every unit is ten times the size of the next smaller unit; for example there are 10 millimeters in 1 centimeter. The familiar English system of measurements is not a decimal system—there are 4 quarts in 1 gallon, 12 inches in 1 foot, and 16 ounces in 1 pound.

The SI system uses a small number of base units (such as the meter) and a set of prefixes (such as kilo-, centi-, and milli-) which can be used with any base unit. Chemists continue to use a few older metric units such as the liter and the milliliter that are not part of the SI system. A volume of 1 liter (1 L) equals 1 cubic decimeter (1 dm^3) and 1 milliliter (1 mL) is equal to 1 cubic centimeter (1 cm^3).

Measured quantities almost always consist of two parts: a number and a unit. You must get in the habit of writing units with your measured numbers. Don't write a length as 5; write it as 5 mm, 5 cm, or 5 m.

MEASUREMENTS ARE ALWAYS ESTIMATES

Much of the work in a chemistry laboratory involves measurements. Unlike counting, which can be exact, measurements are never exact but are always estimated quantities. All measurements are comparisons to some standard. No matter how precise the measurement instrument, the last digit measured must be an estimate. Obviously, some instruments make better estimates than others and experienced investigators make better measurements than inexperienced investigators.

In making length measurements with a 12-inch ruler, the familiar English system divides each into halves, quarters, eighths, and sixteenths of an inch. Because it is not a decimal number system, measurements with the English ruler are usually done by finding the mark closest to the length being measured. In Figure 2-1, the length is estimated as $1\frac{7}{16}$ inch.

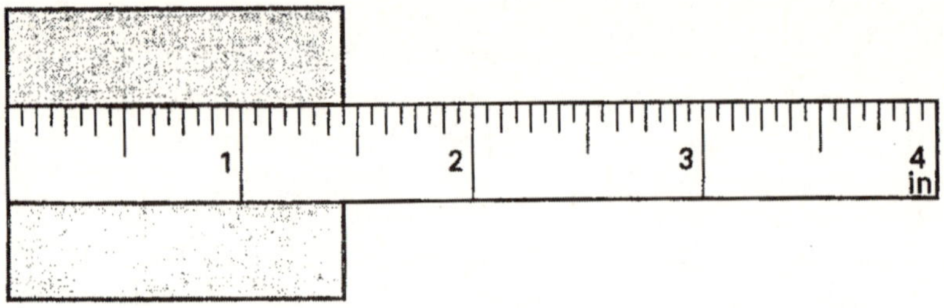

Figure 2-1 Measure to the nearest mark ($1\frac{7}{16}$ inch) when using an English ruler.

In contrast, because the SI system is a decimal system, each centimeter is divided into tenths of a centimeter rather than sixteenths of a centimeter. A length of 0.1 cm is the same as 1 mm.

Most measurements are read from a scale such as the one in Figure 2-2. A scale such as this is commonly used on instruments that measure length (a ruler), mass (a balance), temperature (a thermometer), and volume (volumetric glassware such as a graduated cylinder, a buret, or a syringe). The scale has graduation marks and some of the graduation marks have printed numbers. Usually there are 10 graduation marks between the printed numbers but some scales have only 5 graduation marks between the printed numbers.

In using the scale, you should imagine additional imaginary graduation marks that are between the marks printed on the scale. As a rule of thumb, you should try to imagine 10 additional marks between the printed marks. The length in Figure 2-2 is between two graduation marks, 4 cm and 5 cm. The 10 imaginary marks are 4.0, 4.1, 4.2, 4.3, 4.4, 4.5, 4.6, 4.7, 4.8, and 4.9 cm. The length can be estimated as 4.2 cm.

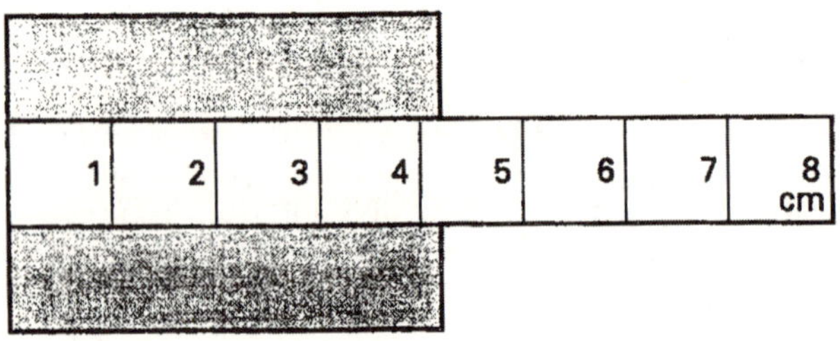

Figure 2-2 The reading (4.2 cm) is between two graduation marks.

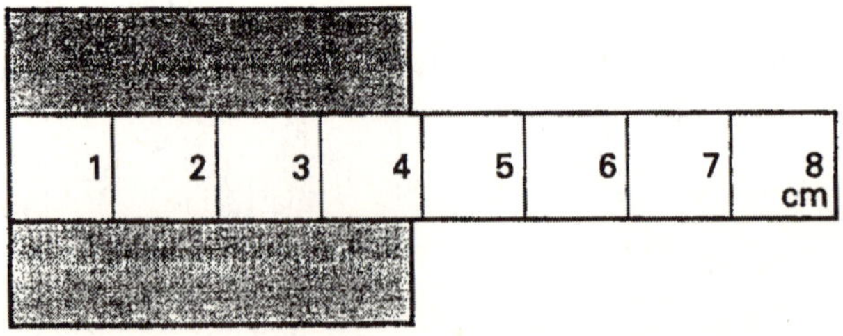

Figure 2-3 The reading is 3.9. Estimate using imaginary marks between two graduation marks.

When the reading appears to lie exactly on a printed graduation mark, add a zero to the reading to indicate the imaginary mark was zero. The length in Figure 2-4 should be estimated as 4.0 cm.

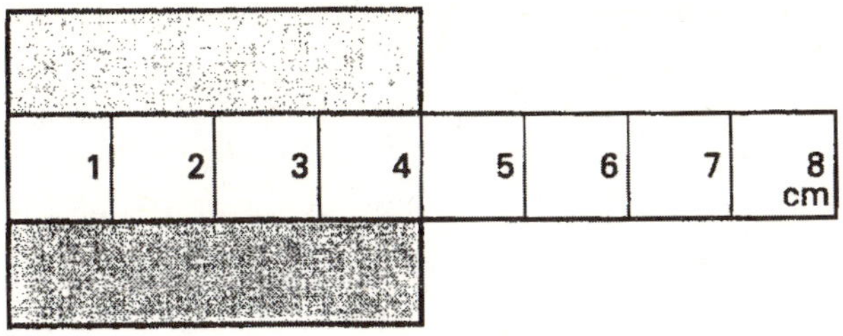

Figure 2-4 The reading (4.0 cm) is on a printed graduation mark.

Each of these last three measurements used the same measurement instrument, one where the smallest graduation mark was 1 cm. Each measurement was estimated to the nearest 0.1 cm. Thus, each measurement has 1 digit after the decimal point. The measurement in Figure 2-4 is 4.0 cm rather than 4 cm. The measurement of 4.0 cm indicates the measurement is estimated ±0.1 cm whereas a measurement of 4 cm indicates a less precise estimate of ±1 cm.

When the printed marks are very close together it may not be possible to imagine 10 additional marks and you may be able to imagine only 1 additional mark. An investigator can decide whether the reading appears to lie ON a graduation mark or BETWEEN two closely spaced graduation marks. The graduation marks in Figure 2-5 are every 0.1 cm (1 mm). Although it is possible to imagine 10 additional marks, inexperienced investigators might imagine only 1 additional mark. The length in Figure 2-5 is BETWEEN 4.0 cm and 4.1 cm. If it were ON the 4.0 cm mark we could write the measurement as 4.00 cm. If it were ON the 4.1 cm mark we could write the measurement as 4.10 cm. Since the length appears to be BETWEEN 4.0 and 4.1 cm, we could use an imaginary mark to get the measurement 4.05 cm. This measurement has been estimated to the nearest 0.05 cm. With experience it is possible to make better estimates for the imaginary marks.

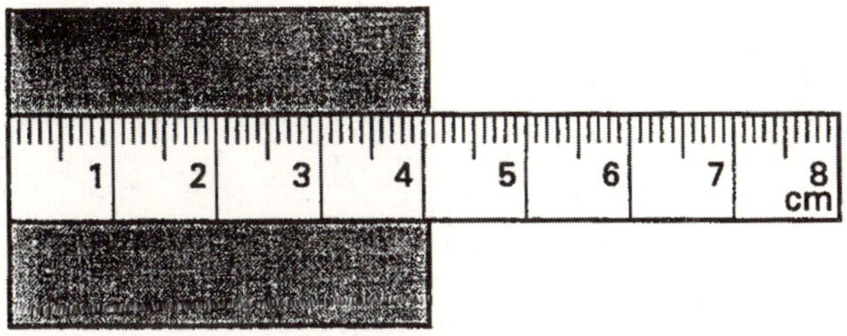

Figure 2-5 When the printed marks are very close together, you can use 1 intermediate imaginary mark. The reading is 4.05 cm.

Generally, no matter how closely spaced the graduation marks, you should estimate the measurement by imagining additional marks between the printed marks on the scale. This measurement estimation technique should be used for all measurement scales whether found on a ruler, balance, thermometer, or graduated cylinder.

USE OF THE THERMOMETER

Review Section 9 of Fundamental Laboratory Operations, page 24, for procedures in using a laboratory thermometer.

USE OF THE BALANCE

Review Section 4 of Fundamental Laboratory Operations, page 15, for general procedures in using a laboratory balance. Your instructor will supply specific directions for use of the balance in your laboratory. Remember that this measuring instrument is a balance, not a scale.

VOLUMETRIC GLASSWARE

Be thoroughly familiar with Section 3 of Fundamental Laboratory Operations, pages 10-15, for descriptions and proper uses of graduated cylinders and volumetric flasks. All glassware must be thoroughly cleaned before use and cleaned again before storage.

DENSITY

Density is the mass of a substance per unit volume of that substance. In equation form:

$$\text{density} = \frac{\text{mass}}{\text{volume}} \quad \text{or} \quad d = \frac{m}{v}$$

The density of a solid or liquid is expressed in units of grams per cubic centimeter (g/cm^3) or the equivalent grams per milliliter (g/mL). Density is a constant at a particular temperature and does not depend on the size of the sample. Water, for example, has a density of 1.00 g/mL at 4°C. At any other temperature liquid water is slightly less dense, but still very close to 1.00 g/mL.

When determining the density of a liquid, mass and volume can be obtained by direct measurements; a balance is used for measuring mass, and a graduated cylinder is used for measuring volume.

If a solid is insoluble in water, its volume can be indirectly determined by measuring its water displacement. Water is poured into a graduated cylinder and its volume carefully measured. The graduated cylinder is tilted and the solid is allowed to submerge beneath the surface of the water. The change in volume is the volume of the solid. See Figure 2-6.

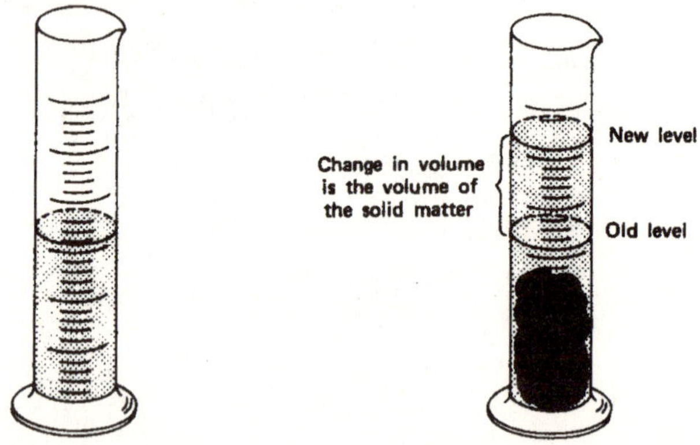

Figure 2-6 Measuring the volume of a solid by water displacement.

If a solid object floats on water, its density must be less than water's density. If the solid's density is greater than water's density, the solid will sink in water.

SPECIFIC GRAVITY

The specific gravity of a liquid or solid is the density of that substance divided by the density of water (measured at the same temperature).

$$\text{sp. gr.} = \frac{\text{density of the substance}}{\text{density of water}}$$

Recall that water's density is 1.00 g/mL at 4°C. If the density of the substance is also expressed in units of g/mL, the specific gravity of the substance will be the same number as the density but without any units—the units cancel. For example, ethyl alcohol has a density of 0.789 g/mL and a specific gravity of 0.789. Specific gravity is one of very few quantities without any units.

The specific gravity of urine is valuable in medical diagnosis. The specific gravity indicates the concentration of solids such as urea, phosphates, chlorides, proteins, and sugars which are dissolved in the urine. The normal specific gravity of urine ranges from 1.005 and 1.030, with most samples falling between 1.010 and 1.025.

If a person's urine produces a continuously low specific gravity reading, this may indicate diabetes insipidus, a disease caused by impairment of an antidiuretic hormone. Low specific gravity readings may also indicate that a kidney has lost its ability to concentrate urine because of tubular damage.

High urine specific gravity readings occur in patients with diabetes mellitus, adrenal inefficiency, hepatic diseases, and congestive cardiac failure. The specific gravity of urine may be elevated due to excessive loss of water caused by sweating, fever, vomiting, or diarrhea.

The specific gravity of a liquid can be measured with a hydrometer. The **hydrometer** consists of two pieces—a jar to hold the liquid, and a float. The float is weighted to make it partially sink in the liquid. The amount the float sinks is determined by the liquid; the float sinks less in liquids with a high specific gravity. The float has a thin stem with a calibrated scale. The specific gravity reading is made at the position the surface of the liquid meets the scale. A **urinometer** is a hydrometer especially designed and calibrated for urine. See Figure 2-7.

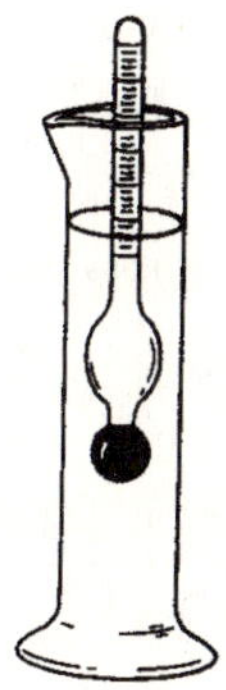

Figure 2-7 A urinometer. The reading is made where the float meets the surface of the liquid.

Typical urinometer scales are calibrated from 1.000 to 1.060. Because of a lack of space on the thin stem of the float, decimal points are often left out; a reading of 1028 is really 1.028. The graduation marks on a urinometer are very close together; so close that it is impractical to use imaginary marks between the printed marks. See Figure 2-8.

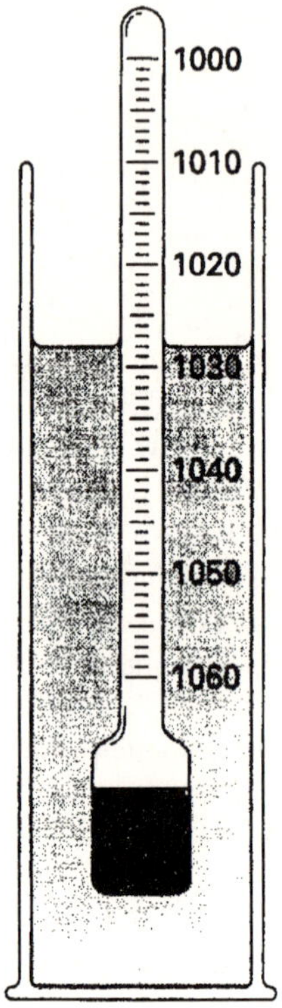

Figure 2-8 The urinometer reading is 1.028. Notice that the numbers increase going ***down*** the stem of the float.

For accurate work, the temperature of the liquid should be taken. A small correction to the specific gravity can be applied for every degree the liquid is above or below 15.6°C. This experiment will not make a temperature correction.

Procedure

I. LENGTH

Use an English-metric ruler to make the following measurements as accurately as you can. *Make sure you write the units of each measurement along with the number.* Use a pen or pencil to check off each step as you complete the step.

1. Measure the length of the smallest test tube in your locker in milliliters and in inches.

2. Measure the height of a 250-mL beaker in centimeters and inches.

3. Measure the dimensions of the front cover of this lab manual in English units and in SI units. Calculate its area using each set of units.

II. TEMPERATURE

1. Measure the temperature of the air in the room.

2. Measure the temperature of the air outdoors.

3. Fill a beaker with cold tap water. Measure the temperature of the cold tap water.

4. Fill a beaker with hot tap water. Measure the temperature of the hot tap water.

III. MASS

1. Measure the mass of a 250-mL beaker.

2. Measure the mass of a piece of weighing paper.

   [If your balance has a "tare" function, you may use the following procedure: Place the paper on the pan of the balance and, if you choose, press "tare." On the Report Sheet for the "mass of the weighing paper" you may write "TARE" or "0.00 g" rather than the mass.]

3. Place 5 aspirin tablets on the weighing paper and record the mass of the paper and tablets.

   [If you have chosen to use the "tare" function of your balance, write "TARE" or "0.00 g" for the "mass of the weighing paper and 5 aspirin tablets." Record the "mass of 5 aspirin tablets" and proceed to Step 5.]

4. Calculate the mass of the 5 aspirin tablets.

5. Calculate the average mass of 1 aspirin tablet.

6. Follow your instructor's directions for the disposal of the aspirin tablets.

IV. VOLUME

1. Examine the graduation marks of a 50-mL graduated cylinder. Look carefully, some graduation marks may be missing. On the Report Sheet draw each graduation mark below the 10-mL mark. Label all the graduation marks with their volume.

2. Examine the graduation marks of a 10-mL graduated cylinder. On the Report Sheet draw each graduation mark below the 3-mL mark. Label all the graduation marks with their volume.

3. Fill a standard (16 × 125 mm) test tube to the brim with tap water and pour the contents into a 50-mL graduated cylinder. Measure the volume to the nearest 0.1 mL. Remember to read the graduated cylinder at the bottom of the meniscus.

4. Fill a small (10 × 75 mm) test tube to the brim with tap water and pour the contents into a 10-mL graduated cylinder. Measure the volume to the nearest 0.1 mL.

5. Half fill a small beaker with tap water. Use a medicine dropper to transfer 1 mL of water to a 10-mL graduated cylinder. Count the number of drops in 1 mL.

6. Obtain a rectangular container from your instructor. Measure the inner dimensions in centimeters. Calculate its volume in cm^3.

7. Fill the rectangular container to the brim with tap water. Make sure the water is at the brim, not above or below it. Carefully pour the contents into a graduated cylinder. A funnel can be used to avoid spillage. Record the volume in the cylinder.

V. DENSITY

A. Density of Water

1. Measure the mass of a dry 50-mL graduated cylinder.

2. Pour exactly 50.0 mL of deionized water into the graduated cylinder. Use a medicine dropper to adjust the meniscus to the 50-mL mark.

3. Measure the mass of the graduated cylinder with the water.

4. Calculate the mass of the water.

5. Calculate the density of the water. Remember to include units with each quantity.

B. Density of a Sugar-Water Solution

1. Measure the mass of a clean, dry 100-mL volumetric flask with its cap.

2. Pour about two teaspoons of sugar into the flask. Use a cone-shaped piece of weighing paper as a funnel.

3. Fill the flask halfway with deionized water. Swirl the flask until the sugar dissolves.

4. Fill the flask to the 100-mL mark with deionized water. Use a medicine dropper to carefully add the last milliliter of water. Remember that the bottom of the meniscus should be at the 100-mL mark.

5. Tightly cap the flask and invert it about 15–20 times to obtain a uniform solution. Measure the mass of the capped flask and its contents.

6. Calculate the mass of the sugar-water solution.

7. Calculate the density of the sugar-water solution.

8. Save the solution for use in Part VI.

C. Density of a Rubber Stopper

1. Measure the mass of a No. 1 solid rubber stopper.

2. Pour 20–25 mL of tap water into a 50-mL graduated cylinder. Record the volume to the nearest 0.1 mL.

3. Tip the graduated cylinder and slowly slide the stopper into the graduated cylinder until it is entirely beneath the surface of the water. Place the graduated cylinder in an upright position and record the volume to the nearest 0.1 mL.

4. Calculate the volume of the rubber stopper and the density of the rubber stopper.

D. Density of an Unknown Solid

Obtain an unknown solid from your instructor and determine its density.

VI. SPECIFIC GRAVITY

1. Use a urinometer to measure the specific gravity of deionized water. Fill the urinometer jar with deionized water to 2 cm from the top.

2. Gently add the float to the water with a slight spinning motion. Avoid wetting the float stem above the liquid line. Make sure the float is actually floating and not resting on the bottom of the jar.

3. Read the scale of the urinometer float at the lowest portion of the meniscus of the water. Keep the float away from the sides of the jar while reading.

4. Use the urinometer to measure the specific gravity of the sugar-water solution prepared in Part V. B.

5. Discard the sugar-water solution in the sink. Rinse the volumetric flask and the urinometer several times with tap water and once or twice with a few milliliters of deionized water.

6. Wash your hands before leaving the laboratory.

Name ______________________ Section ____________ Date ______________

Prelab Questions for Experiment 2

1. Record the following measurements.

a. 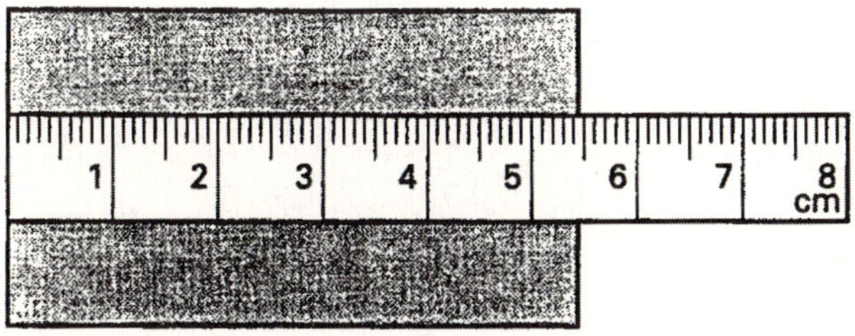

b. 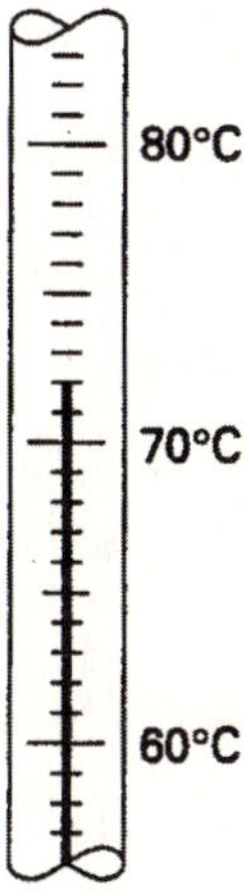

c. 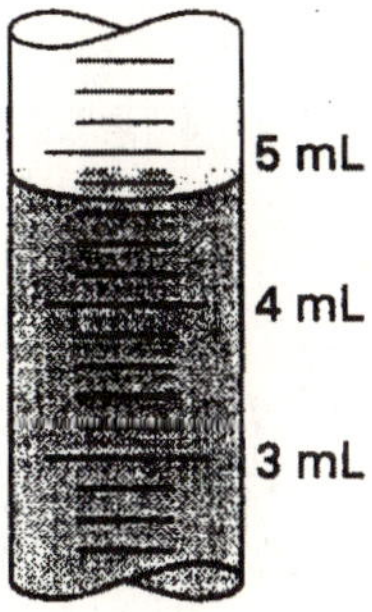

d. 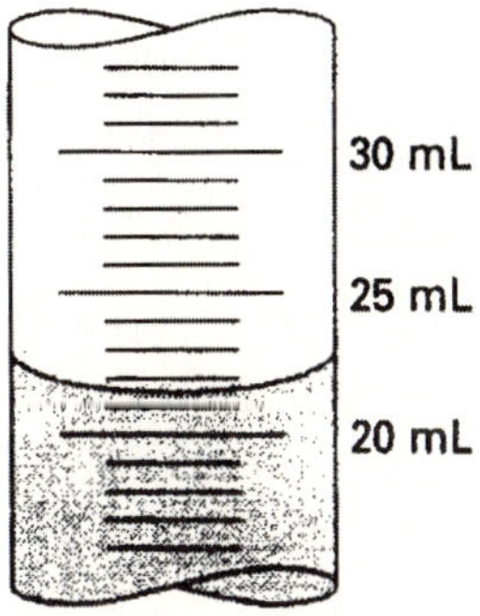

2. A 15-gram solid object is dropped into a graduated cylinder containing 35 mL of water. The water level rises to the 45-mL mark.

   a. What is the density of the solid? Show your calculations.

   b. What is the specific gravity of the solid?

3. a. What is a meniscus?

   ____

   b. Explain the proper technique for reading a meniscus.

   ____

   ____

4. When preparing a solution with a volumetric flask, why is the solute first dissolved in a small quantity of solvent before filling the flask to the mark?

   ____

   ____

Name ______________________ Section ____________ Date ______________

Report for Experiment 2

I. LENGTH

Remember to include units with each quantity.

1. a. Length of a small test tube ____________ mm ____________ in

   b. Why is the small test tube called a 10 × 75-mm test tube?

   ____

2. a. Height of 250-mL beaker ____________ cm ____________ in

3. a. Dimensions of front cover in English units ____________ × ____________

   b. Dimensions of front cover in SI units ____________ × ____________

   c. Area of front cover in English units ____________

   d. Area of front cover in SI units ____________

II. TEMPERATURE

1. Temperature of inside air ____________
2. Temperature of outside air ____________
3. Temperature of cold tap water ____________
4. Temperature of hot tap water ____________

III. MASS

1. Mass of 250-mL beaker ____________
2. Mass of weighing paper and 5 aspirin tablets ____________
3. Mass of weighing paper ____________
4. Mass of 5 aspirin tablets ____________
5. a. Average mass of 1 aspirin tablet

   b. If the mass of aspirin in each tablet is 5 grains (1 grain = 0.065 g), calculate the mass in grams of aspirin in 1 tablet.

Calculations: ____________

c. Calculate the percentage of aspirin in each tablet.

$$\% \text{ aspirin} = \frac{\text{mass of aspirin}}{\text{mass of tablet}} \times 100$$

IV. VOLUME

1. Draw and label the graduation marks on a 50-mL graduated cylinder below the 10-mL mark.

2. Draw and label the graduation marks on a 10-mL graduated cylinder below the 3-mL mark.

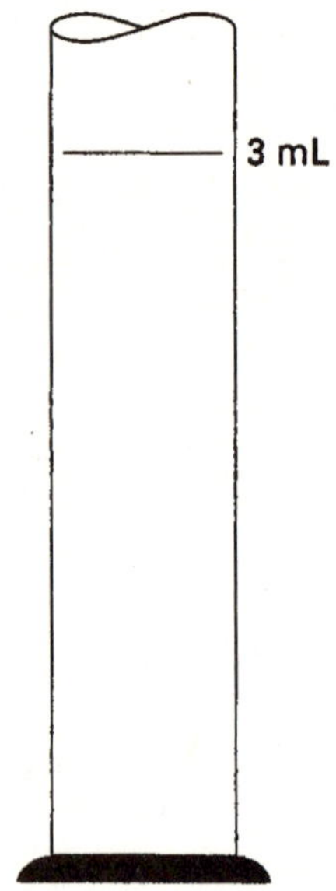

3. Volume of standard test tube ____________

4. Volume of small test tube ____________

5. Number of drops in 1 mL ____________

6. Rectangular container ____________

   a. Inner dimensions ____________ × ____________ × ____________

   b. Volume calculated from dimensions

   c. Volume of water to fill container

   d. Does your data indicate that 1 cm^3 is equal to 1 mL? ____________

Name ______________________ Section ____________ Date ______________

V. DENSITY

A. Density of Water

1. Mass of empty graduated cylinder ____________
2. Mass of graduated cylinder and water ____________
3. Mass of water ____________
4. Volume of water ____________
5. Density of water ____________

   Calculations:

6. Why might the experimentally determined density of water differ from its expected density?

   ____

B. Density of Sugar-Water Solution

1. Mass of volumetric flask and solution ____________
2. Mass of empty volumetric flask ____________
3. Mass of solution ____________
4. Density of solution ____________

   Calculations:

C. Density of a Rubber Stopper

1. Mass of rubber stopper ____________
2. Volume of water and rubber stopper ____________
3. Volume of water ____________
4. Volume of rubber stopper ____________
5. Density of rubber stopper ____________

   Calculations:

D. Density of an Unknown Solid

Unknown Number ______________

1. Mass of solid ______________
2. Volume of water and solid ______________
3. Volume of water ______________
4. Volume of solid ______________
5. Density of solid ______________

Calculations:

VI. SPECIFIC GRAVITY

1. a. Specific gravity of water ______________

   b. How does the specific gravity of water measured with the urinometer compare with the density calculated in Part V. A?

   ____

   ____

2. a. Specific gravity of sugar-water solution ______________

   b. How does the specific gravity of the sugar-water solution measured with the urinometer compare with the density calculated in Part V. B?

   ____

   ____

   c. If your sugar-water solution were a urine specimen, what would the specific gravity tell you about the condition of the patient?

   ____

   ____

Name ______________________ Section ____________ Date ____________

Related Questions for Experiment 2

1. Some brands of pain relievers contain aspirin plus other active ingredients such as acetominophen, buffers, and caffeine. Does the label of the aspirin bottle used in this experiment list any other active ingredients?

____

2. What are the inactive ingredients in aspirin? Hint: Use the Physician's Desk Reference to look up the "Inactive Ingredients" in two brands of aspirin.

Brand and ingredients: ____

____

Brand and ingredients: ____

____

3. What is the purpose of the inactive ingredients in aspirin?

____

____

4. a. Could you determine the volume of a cube of sugar using the water displacement method? ____________

   b. Why or why not?

____

Experiment 3

Paper and Thin-Layer Chromatography

Purpose: To become familiar with two chromatographic methods of analysis by identifying the components in black ink, separating the components of food coloring, and isolating the pigments found in spinach.

Introduction

Most samples of matter are impure mixtures of two or more substances. Separating and identifying the components of a mixture are important goals in chemistry. One common and powerful separation technique is **chromatography**. The word chromatography means "separation of colors" but today chromatography is used for both colored and colorless substances. The separation process is based on the fact that a porous solid **adsorbs*** different substances to different extremes. The components of the mixture become separated between the stationary porous solid phase and the mobile liquid phase, which passes through the stationary phase. As a liquid mobile phase moves by capillary action, it "pulls" along various molecules that have been placed on the solid stationary phase. Different types of molecules have different attractions to the mobile and to the stationary phases, and therefore do not travel at the same speed through the stationary phase. This leads to a separation of the various molecules.

The simplest types of chromatography, paper and thin-layer, will be used in this experiment. Other chromatographic methods, including column chromatography, gas chromatography (gc), and high-performance liquid chromatography (hplc) are used extensively in chemistry and related fields such as medicine.

In medicine, chromatography is used to separate and identify amino acids and proteins in mixtures. Chromatograms of blood samples will sometimes reveal the presence of foreign proteins associated with certain diseases. Law enforcement agencies sometimes require chromatographic analysis of urine specimens from suspected drug addicts.

PAPER CHROMATOGRAPHY

In paper chromatography the stationary phase is a sheet of absorbent paper, such as filter paper. A tiny drop of the mixture to be separated is placed on the paper near the bottom of the paper. A lightly drawn pencil line marks the location of the spot. This location is called the **origin**. The paper is suspended vertically in the mobile phase, a solvent or **eluent**. The eluent could be water, another pure liquid such as alcohol, or a solution. Different eluents will differ in their ability to "pull" along the substances in the mixture being separated.

***Adsorption** is the adhesion of a substance to the *surface* of another substance, as opposed to **absorption**, in which a substance penetrates into the inner structure of another substance.

The origin must be above the surface of the eluent. The eluent rises up the paper by capillary action. When the eluent reaches the origin, the components of the mixture rise at different rates. The container must be covered to prevent evaporation of eluent. The chromatogram must be removed from the eluent before the eluent reaches the top of the paper.

As the substances in the mixture rise up the paper, they spread out and the spots become larger. For this reason, the original spot should be as small as possible, less than 5 mm in diameter. If too much material is applied to the small spot or the sample is too concentrated, the spot may develop a long "tail." If too little material is applied to the spot, the color of the spot may be too faint to see as the spot enlarges while moving up the paper. Trial-and-error and experience help the experimenter obtain both a small spot and one with the proper amount of material.

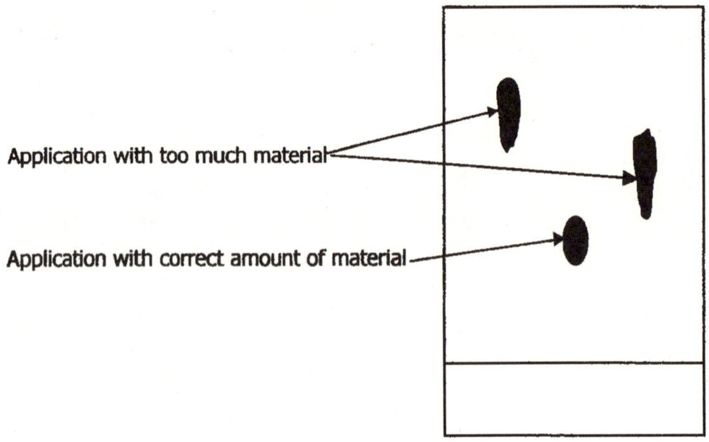

Figure 3-1 Chromatogram "tailing."

Substances can be identified by the heights they reach on the completed chromatography by calculating Rf (rate of flow) values. The Rf value is a constant for a given substance under the same experimental conditions. The Rf value may be calculated from the following equation.

$$R_f = \frac{\text{distance traveled by the substance from the origin}}{\text{distance traveled by the eluent from the origin}}$$

Figure 3-2 shows the finished chromatogram of substance A, substance B, and a mixture containing substances A and B. For each substance, measure the distance from the origin to the center of the migrated spot. If the spot is large with a "tail," measure to the "center of gravity" of the spot.

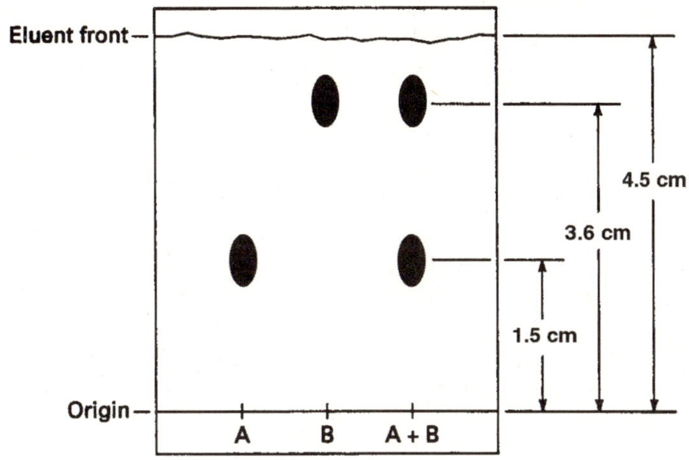

Figure 3-2 Typical finished chromatogram of two substances.

$$R_f \text{ (substance A)} = \frac{1.5 \text{ cm}}{4.5 \text{ cm}} = 0.33$$

$$R_f \text{ (substance B)} = \frac{3.6 \text{ cm}}{4.5 \text{ cm}} = 0.80$$

Once the Rf value is known, the substance can sometimes be identified by comparing its Rf value with those reported in the literature. To check the identity of an unknown substance, it is usually necessary to run a chromatogram of a known sample simultaneously with the unknown.

THIN-LAYER CHROMATOGRAPHY (TLC)

Thin-layer chromatography is almost identical to paper chromatography. Instead of using paper, the stationary phase is a thin coating of adsorbent material, called the **sorbent**, on a sheet of glass, plastic, or metal. As in paper chromatography, the TLC sheet is suspended vertically in an eluent and the eluent travels up the sheet. TLC offers two advantages over paper chromatography. First, it provides a better separation of the mixture with less spreading of the spots; second, the sorbent may be varied. Common sorbents include silica (SiO_2, very pure, finely ground sand), alumina (Al_2O_3, also used in abrasives, ceramic materials, and dental cement), and cellulose (similar to very pure, finely ground wood fibers). In order to separate the substances of a mixture, the substances must have different R_f values. By carefully choosing an eluent and a sorbent, it is usually possible to find a combination that will separate the mixture.

FORENSIC CHEMISTRY: BLACK INK PIGMENTS

Do all black pens contain the same pigments? In this simulation, the police have asked your assistance in solving a crime. One of several suspects has written a ransom note. Since each suspect uses a different brand black felt-tipped pen, chromatography can identify which pen wrote the ransom note. This is one example of the use of chemistry in **forensic science**—the use of science in legal proceedings.

In a real case, forensic chemists could use a solvent to remove the black ink from a small portion of the ransom note. Then the ink could be analyzed by chromatography and compared to known inks. In our simulation your instructor will use a pen to apply an "unknown" ink spot to your chromatography paper. By comparison with known inks you will identify the unknown ink.

SEPARATION OF SPINACH PIGMENTS

Deeply colored vegetables such as spinach contain a mixture of pigments including chlorophyll-a (blue-green), chlorophyll-b (yellow-green), and b-carotene (yellow-orange). When exposed to air the chlorophyll pigments are slowly oxidized to form brown-colored products. The pigments are nonpolar and do not dissolve in water, a highly polar solvent; that's why grass stains are so difficult to launder from clothing. The pigments do dissolve in acetone, a common solvent found in fingernail polish remover. The eluent for the chromatography of these pigments will be a 2:1 mixture of ligroin and acetone. Ligroin is a non-polar solvent similar to gasoline, mineral spirits, or painter's naphtha—it is a mixture of hydrocarbons with a boiling point range of 130–145°C. The 2:1 ligroin-acetone eluent mixture is chosen because its polarity gives a good separation of the spinach pigments. The eluent mixture must be free of water—one drop of water would considerably change the polarity of the mixture.

FOOD COLORING

Color greatly influences our perceptions about the world around us—including our judgments about the quality and appeal of the products we buy and use. In a 1970s research project, volunteers ate part of a meal under special lighting that concealed that the colors of the foods had been altered. When, under normal lighting, the diners discovered that their steaks were blue, peas red, and French fries green, some participants became ill at the sight of the unnaturally colored food they had been eating.

Obviously, color is important in consumer acceptance of a product. Even the color of the container can make a difference in consumer purchases. For these reasons people have been coloring foods, drugs, and cosmetic products for thousands of years. Food was once colored with dyes such as those from beets, peppers, grape skins, saffron, and insect bodies. In the 19th century as the country changed from an agricultural to an industrialized nation, colorants containing lead, copper, arsenic, or other toxic substances became more widely used and presented significant health risks. In response, the Federal Food and Drug Act of 1906 and the Federal Food, Drug and Cosmetic Act of 1938 required government certification and control over color additives for drugs and cosmetics as well as foods. A 1960 amendment to the FD&C Act included the Delaney Clause which banned additives shown to induce cancer in humans or animals, even at very low doses. The Delaney Clause was based on the premise that there is no safe threshold for cancer-causing substances.

The U.S. Food and Drug Administration banned Red Dye No. 2 in 1976 because it was suspected, but not proven, to cause cancer. The FDA proposed banning the artificial sweetener saccharine in 1977 because studies showed it could cause cancer. Public outcry over the loss of this sweetener forced Congress to specifically exempt saccharine from the Delaney Clause. Interestingly, based on the same research studies, Canada prohibits most uses of saccharine and permits use of the sweetener cyclamate and Red Dye No. 2, exactly the opposite of the U.S.

One of the seven FD&C approved food colorants, Red No. 3, has been shown to cause thyroid cancer in male rats. Food manufacturers and the cosmetic industry fought off more than a dozen attempts to prohibit Red No. 3. Finally the FDA banned one form of Red Dye No. 3; Red No. 3 Lake pigments previously used in products such as lipsticks, candies, and pill coatings were banned in 1990. Lake pigments are an insoluble version of the colorant. Unlike a lake, a dye is soluble in water. The dye version of Red 3 is still permitted in many drugs and foods such as baked goods, dairy products, desserts, jellies, vegetable products, gelatin desserts, pistachio nuts, and powdered beverage mixes. One clearly visible use is in maraschino cherries. Red No. 3 is the only dye that doesn't bleed in citrus juices. Without it, a fruit-cocktail cherry would be brown.

Many have argued that the risk of human cancer from Red No. 3 is extremely small. The Delaney Clause requires the FDA to ban additives if there is any cancer risk, no matter how small. Some have suggested that the Delaney Clause be replaced by a standard that allows for a small, insignificant amount of cancer risk, such as one cancer case in 1 million people. Others have argued that while there are some cancer risks that are unavoidable, we ought to eliminate all the avoidable risks. Food colorants are clearly avoidable because they are intentionally added. If the zero-risk Delaney Clause remains in force, the FDA is someday likely to totally ban Red 3.

Seven "certified" synthetic FD&C dyes can be added to food products: Red 3, Red 40, Blue 1, Blue 2, Yellow 6, and Green 3. Since Yellow 5 causes severe allergic reactions in some people (about 1 in 10,000), the FDA requires its listing on food labels. Yellow 5 was once blamed for hyperactivity in some children but now it is clear that most hyperactivity is not caused by this or other additives. The other six food dyes do not have to be separately identified on food labels and can be listed as "artificial colors." The average U.S. citizen annually consumes about 3 grams of these dyes. Twenty-six other color additives are exempt from certification and are permitted in food—these include caramel, grape skin extract, paprika, turmeric, iron extract, titanium dioxide, and b-carotene. Most exempt colorants are derived from vegetable or mineral sources and are therefore often called "natural"; they can also be made synthetically.

Out of 200 colorants once approved in the U.S., only these seven certified and 26 exempt colorants remain approved for addition to our food.

This experiment will examine food colors and determine why grape-flavored Kool-Aid® has a purple color.

Procedure

I. PAPER CHROMATOGRAPHY: FOOD COLORING

CAUTION: Acetone and ligroin are flammable. Do not use near open flames.

Pregnant women may be advised by their physicians to avoid breathing the vapors of solvents such as acetone and ligroin.

1. Cut a 46- × 56-cm piece of chromatography paper into four strips, each 14- × 46-cm. Cut a 9- × 14-cm piece from one of the strips.

2. Use a pencil to draw the origin line about 1.5 cm from the bottom of the 14-cm edge of the paper. Use a pencil to make marks every 1.5 cm along the length of the origin line, starting about 2.5 cm from the edge of the paper. See Figure 3-3.

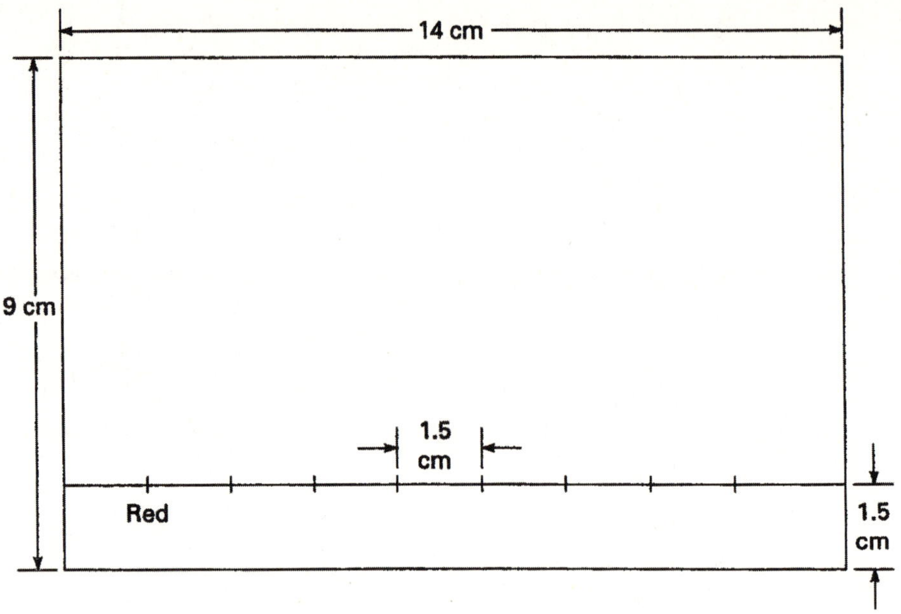

Figure 3-3 Labeling the origin line on the chromatography paper.

3. Place 1 drop of food color on a watch glass or in a small beaker. Soak a wooden toothpick in the food color for several seconds then briefly touch the toothpick to the first mark.

4. Use a pencil to write the name of the food color below the origin line.

5. Repeat Steps 3 and 4 with each other food color. Use a fresh toothpick for each color.

6. Dissolve 1 packet (3.9 g) of unsweetened grape-flavored Kool-Aid® in 5 mL of water. Stir with a glass stirring rod. Use a fresh toothpick to spot the grape Kool-Aid to the chromatography paper. Reapply the Kool-Aid to the same spot 5 or 6 times to get a darker spot. Use a pencil to write the name "grape" below the origin line.

7. Roll the paper into a cylinder with the origin line at the bottom and the dye spots on the outside of the cylinder. Use two fingers of one hand to hold the ends of the paper close together, about 2 mm apart. Staple the top ends and then staple the bottom ends. The ends of the paper should not touch. See Figure 3.4.

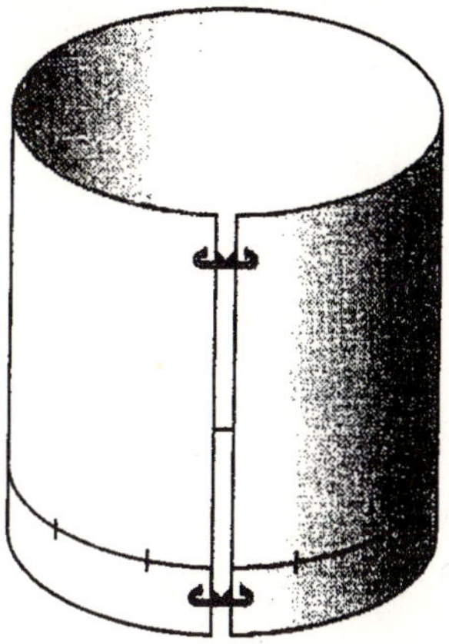

Figure 3-4 Roll into a cylinder and staple with a gap between the ends of the paper.

8. Add about 10 mL of 0.1% salt (NaCl) solution to a 400-mL beaker. The solution should be about 0.5-cm deep.

9. Carefully place the paper cylinder into the beaker containing the salt solution. Make sure the origin line is at the bottom. The origin line must not be below the surface of the eluent. The paper should not touch the side of the beaker.

10. Cover the beaker with a watch glass. Do not disturb the beaker while the eluent rises up the paper.

11. Watch the eluent as it moves up the paper and see what happens as it comes into contact with the food colors. Leave the paper in the beaker until the eluent front is about 1.5 cm from the top of the paper.

12. When the eluent front nears the top, remove the chromatogram and open the cylinder by tearing the paper at the staples. Set the chromatogram on an empty beaker or a paper towel to dry. Because of the large amount of water absorbed by the chromatography paper, the eluent front may continue to move up the paper for several minutes, dragging the dyes with it. After 2 or 3 minutes, or when the paper appears to be drying out, mark the final position of the eluent front with a pencil line.

13. Does each color contain 1 dye or a mixture of dyes? Calculate the R_f value of each dye. Try to identify the name (Red 40, etc.) of each dye. Identify the dyes present in the grape Kool-Aid.

II. TLC: SEPARATION OF SPINACH PIGMENTS

CAUTION: If possible, perform all parts of this procedure using acetone and the acetone-ligroin mixture in a fume hood.

1. Use a mortar and pestle to crush about 4 grams (about a tablespoon) of spinach into very small pieces. The crushing helps break cell walls and free the pigments from the cells.

2. Place the spinach in a clean, dry 150-mL beaker and add 10 mL of acetone. **CAUTION: acetone is flammable. No flames present!** Stir with a glass stirring rod for about 10 minutes.

3. Allow the sediment to settle to the bottom of the beaker.

4. Using a pencil, mark the origin on a 1.5- × 12.5-cm silica gel TLC sheet. Mark the origin lightly at each edge of the silica gel sheet about 1.5 cm from the bottom. Do not use a pen. Do not draw a line completely across the TLC sheet since the sorbent may flake off.

5. Dip a clean capillary tube into the spinach extract. Apply the spinach extract to the origin of the silica gel TLC sheet by quickly touching the capillary tube to the sheet. Hold the capillary tube at right angles to the sheet. Do not scrape off the sorbent with the capillary tube. Allow the spot to dry for 30 seconds and repeat the application about 15 times. The drying period is necessary to ensure a small spot. The repeated application is necessary to ensure sufficient material is applied.

6. Pour 2 mL of the 2:1 ligroin-acetone eluent mixture into a dry 25- × 150-mm test tube (the largest of the three test tube sizes). The 10-mL graduated cylinder used to measure the ligroin-acetone mixture must be dry—we don't want water added to the mixture. **CAUTION: ligroin and acetone are flammable. Keep away from flames.** Stand the test tube in a 250-mL Erlenmeyer flask.

7. Lower the TLC sheet into the test tube and stopper the test tube. The origin on the TLC sheet must be above the surface of the ligroin-acetone eluent mixture. Do not disturb the test tube until the eluent front is 1 cm from the top of the TLC sheet. It may take 40–50 minutes for the eluent front to reach that point.

8. Rinse the graduated cylinder with a small amount of acetone and pour into the "Waste Organic Solvent" container. Wash the graduated cylinder with a brush and laboratory soap solution or the detergent designated by your instructor; rinsing with water will not remove non-polar substances such as ligroin. Go on to Part III while waiting for the spinach chromatogram to develop.

9. Use forceps to remove the TLC sheet when the eluent is 1 cm from the top of the TLC sheet. Do not allow the eluent to reach the top of the sheet. Immediately mark the position of the eluent front. This eluent evaporates very rapidly and soon you will no longer be able to see the position of the eluent front.

10. Allow the sheet to dry and circle all visible spots. The spots may fade or change colors after exposure to air and light.

11. Beginning at the origin, label all spots as A, B, C, D, etc. and calculate their R_f values.

12. Try to identify the pigments by their colors. Chlorophyll-a is blue-green, chlorophyll-b is yellow-green, and β-carotene is yellow-orange.

13. **Disposal:** Place the excess ligroin-acetone eluent mixture and the excess acetone-spinach extract in the "Waste Organic Solvent" container. Place the excess spinach in the "Waste Spinach" container.

III. PAPER CHROMATOGRAPHY: FORENSIC CHEMISTRY

1. As in Part I, cut a strip of chromatography paper about 9 × 14 cm.

2. Using a pencil, draw the origin line 1.5 cm from the bottom of the paper. Use a pencil to make marks every 1.5 cm along the origin line.

3. Use a pencil to number the marks below the origin line.

4. Select one of the black pens and touch it briefly to the first numbered mark. Make a black spot about 1 or 2 mm in diameter. Write the name of the pen on the Report Sheet for Spot Number 1.

5. Repeat Step 4 for each pen to be tested.

6. After all known pens have been spotted, give the paper to your instructor who will spot one or more unknown pens. Write the unknown number(s) on the Report Sheet.

7. Roll the paper into a cylinder and staple as in Part I.

8. Add about 10 mL of distilled water to a 400-mL beaker. The water should be about 0.5-cm deep.

9. Carefully place the paper cylinder into the beaker. Make sure the origin line is at the bottom. The origin line must not be below the surface of the water.

10. Cover the beaker with a watch glass. Do not disturb the beaker until the paper is removed.

11. Watch the water as it moves up the strips of paper and see what happens as it comes into contact with the black marks. Leave the paper in the water until the eluent front is about 1.5 cm from the top of the paper.

12. When the eluent front nears the top, remove the chromatogram and open the cylinder. Set the chromatogram on an empty beaker or a paper towel to dry. After 2 or 3 minutes, or when the paper appears to be drying out, mark the final position of the eluent front with a pencil line.

13. Identify the unknown pen(s).

Name ______________________ Section ____________ Date ____________

Prelab Questions for Experiment 3

1. What safety precautions are necessary when using acetone and ligroin?

____

____

2. Define the following terms:

   a. Sorbent ____

   b. Eluent ____

   c. Eluent front ____

   d. Origin ____

3. Why is a pencil used to mark the origin line and not a ball-point or ink pen?

____

____

4. List two errors that would prevent an accurate R_f determination in the TLC setup illustrated in Figure 3-5. Although different from the procedure in this experiment, it is not wrong to stand a TLC sheet in a beaker.

   a. ____

   b. ____

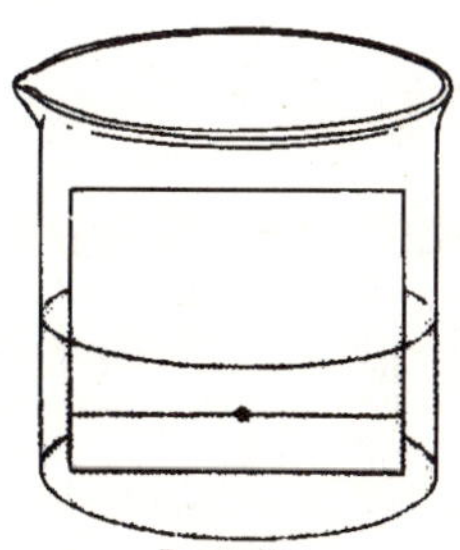

Figure 3-5 A student TLC setup with errors.

5. Calculate the Rf value for each substance on the chromatogram in Figure 3-6.

a. R_f of substance A ______________

Calculations:

b. R_f of substance B ______________

Calculations:

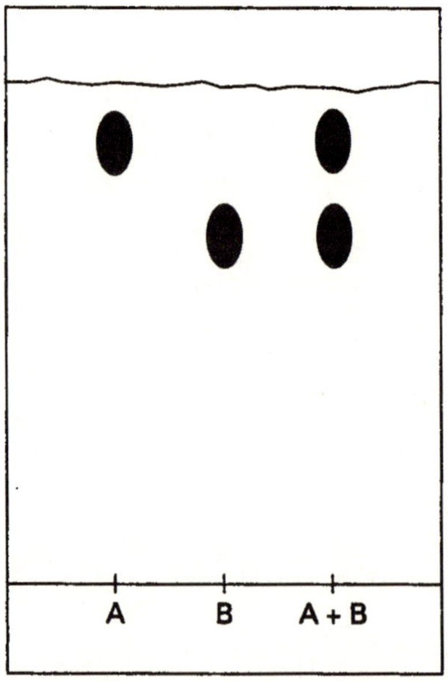

Figure 3-6 A finished chromatogram.

Name ______________________________ Section ____________ Date ______________

Report for Experiment 3

I. PAPER CHROMATOGRAPHY: FOOD COLORING

Yellow Food Color Eluent distance ______________

Color of Spot	Distance Spot Traveled	R_f	Identity of Dye
________	________	________	________
________	________	________	________

Calculations:

Red Food Color Eluent distance ______________

Color of Spot	Distance Spot Traveled	R_f	Identity of Dye
________	________	________	________
________	________	________	________

Calculations:

Blue Food Color Eluent distance ______________

Color of Spot	Distance Spot Traveled	R_f	Identity of Dye
________	________	________	________
________	________	________	________

Calculations:

Green Food Color Eluent distance ____________

Color of Spot	Distance Spot Traveled	R_f	Identity of Dye

Calculations:

Orange Food Color Eluent distance ____________

Color of Spot	Distance Spot Traveled	R_f	Identity of Dye

Calculations:

Grape-flavored Kool-Aid® Eluent distance ____________

Color of Spot	Distance Spot Traveled	R_f	Identity of Dye

Calculations:

Attach your paper chromatogram here.

Name ______________________ Section ____________ Date ______________

II. TLC: SEPARATION OF SPINACH PIGMENTS

Eluent Distance ____________

Spot	Color of Spot	Distance Spot Traveled	R_f	Identity of Pigment
A				
B				
C				
D				
E				
F				

Calculations:

Attach your TLC chromatogram here.

III. PAPER CHROMATOGRAPHY: FORENSIC CHEMISTRY

Spot Number	Name of Pen
1	______________________
2	______________________
3	______________________
4	______________________
5	______________________
6	______________________
7	______________________
8	______________________

Unknown Number __________ Identity of Unknown ______________

Unknown Number __________ Identity of Unknown ______________

Unknown Number __________ Identity of Unknown ______________

Attach your paper chromatogram here.

Name ______________________ Section ____________ Date ____________

Related Questions for Experiment 3

1. The eluent for paper chromatography of food colors was 0.1% NaCl which gives a better separation of dyes than does deionized water. In addition to artificial colors, the spot of grape Kool-Aid contained citric acid, calcium phosphate, salt, maltodextrin, cornstarch, artificial flavor, and ascorbic acid. Since these substances dissolve in water, the eluent moving up the paper was somewhat different for the Kool-Aid than for the food colors. Could this difference have affected the R_f values of the dyes? Is there any evidence to support your answer?

____

____

2. Compound X has an R_f value of 0.25.

   a. How far will compound X have traveled from the origin when the eluent front has traveled 4 cm?

   Calculations: ____________

   b. How far will compound X have traveled from the origin when the eluent front has traveled 8 cm?

   Calculations: ____________

3. Why can an R_f value never be greater than 1.00?

____

____

4. Carrot pigments were separated by TLC chromatography. The chromatogram was removed when the eluent front was only half-way up the TLC sheet.

   a. What is the advantage of allowing the eluent front to rise to near the top of the TLC sheet rather than stopping when only half-way up?

   ____

   ____

   b. Would the R_f values of the pigments differ if the eluent front rose only half-way rather than to near the top of the TLC sheet?

   ____

   ____

5. A student attempted to separate tomato pigments using this experiment's procedure—tomato extract was spotted on a silica gel TLC sheet, developed in a 2:1 ligroin:acetone eluent, and removed when the eluent neared the top of the sheet. Unfortunately, two pigments had the same R_f value and were not separated. What conditions would you change in an effort to retain satisfactory results?

   ____

   ____

6. Look up the entry for FD&C Yellow No. 5 in *The Merck Index*.

   a. What is the common name for FD&C Yellow No. 5? (This is the first name listed.)

   ____

   b. What is the complete chemical name for FD&C Yellow No. 5? (This is the second name listed—this name describes the structure of Yellow No. 5.)

   ____

   ____

   c. What is the molecular formula of Yellow No. 5? ____________

   d. What is the molecular weight (mol wt) of Yellow No. 5? ____________

   e. What uses are listed?

   ____

   ____

Experiment 4
Chemical Periodicity

Reference: *General, Organic and Biological Chemistry: An Integrated Approach, Chapters 2 and 3.*

Purpose: You will study the structure of the periodic table and use the Periodic Law to predict properties of elements, properties of compounds, and formulas of compounds.

Introduction

HISTORIC BACKGROUND

During the eighteenth and nineteenth centuries chemists began to recognize similarities in the properties of some of the elements. They found, for example, that calcium, barium, and strontium could be isolated by similar chemical methods. Sulfur, selenium, and tellurium formed similar chemical compounds, and iodine, bromine, and chlorine were colored nonmetallic elements.

By the early 1800s, the atomic masses (atomic weights) of some of the elements were determined with a fair degree of accuracy. In 1817, Johann Wolfgang Döbereiner made an interesting observation. He noted that in several instances, when three elements having similar chemical and physical properties were arranged in ascending order of atomic masses, the mass of the middle element was almost equal to the average of the first and third. For example consider the similar elements chlorine (atomic mass 35.5), bromine (atomic mass 79.9), and iodine (atomic mass 127). The average of the atomic masses of chlorine and iodine is $(127 + 35.5) \div 2 = 81.3$, fairly close to the accepted atomic mass of bromine (79.9).

Similarly, the atomic mass of strontium, 87.6, is close to the average of the atomic masses of calcium and barium: $(40 + 137) \div 2 = 88.5$. A third such relationship exists in the case of selenium (atomic mass 79.2)—not too different from the average atomic masses of sulfur and tellurium: $(32 + 127.6) \div 2 = 79.7$.

Döbereiner's system of **Triads** was the first systematic attempt to classify the chemical elements but Döbereiner's contemporaries did not pay much attention to his work. Today we realize that Döbereiner's Triads are three elements in the same chemical family. See Figure 4-1.

In 1864, the English chemist John Newlands grouped all the known elements of his day into groupings of seven elements each. He observed that when the elements were arranged in ascending order of atomic mass, the chemical properties of the elements reappeared every eighth element, just as the notes on a musical scale reappear every eighth note. For example, the eighth element after lithium was sodium; lithium and sodium have similar properties. (The noble gases were not discovered until later that century.)

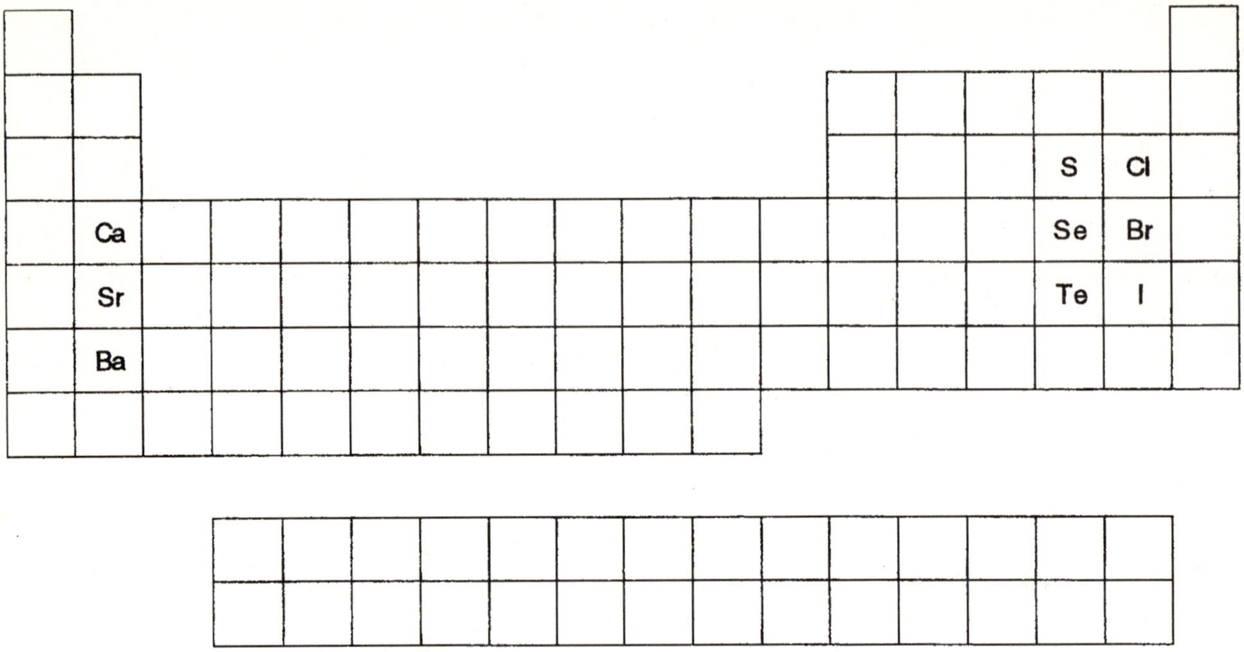

Figure 4-1 Some of Döbereiner's Triads

In a modern version of Newlands' brief table, shown in Figure 4-2, elements are arranged in order of increasing atomic masses. Elements aligned vertically have similar physical and chemical properties. Newlands' **Law of Octaves** didn't work well beyond the element calcium and his work was not well accepted by other chemists who ridiculed his comparison of elements to a piano keyboard. One critic even sarcastically said arranging the elements in alphabetical order would be as useful as in order of atomic mass. His critics were wrong.

Li 6.9 do	Be 9.0 re	B 10.8 me	C 12.0 fa	N 14.0 so	O 16.0 la	F 19.0 ti
Na 23.0 do	Mg 24.3 re	Al 27.0 me	Si 28.1 fa	P 31.0 so	S 32.1 la	Cl 35.5 ti

Figure 4-2 Newlands' Law of Octaves arranged elements in order of increasing atomic masses. Properties repeat themselves like notes on a musical scale.

MENDELEEV'S PERIOD TABLE

The Periodic Table always existed in nature; Newlands and others were groping toward the Period Law but were not yet aware of it. Lothar Meyer, in Germany, and Dmitri Mendeleev, in Russia, independently discovered the Periodic Law and created the periodic table of the elements. Meyer did not publish his work; Mendeleev did in 1869 and received more credit. Unlike Döbereiner and others who sought mathematical relationships among the elements, Mendeleev and Meyer looked for, and found, trends in properties of the elements.

All 63 known elements in Mendeleev's periodic table were arranged in horizontal rows of increasing atomic masses. Because his classification revealed recurring patterns in the elements, Mendeleev was able to leave spaces in his table for elements that he correctly predicted would be discovered. His table gained increased acceptance when three predicted elements were subsequently discovered. The first of these was gallium, discovered in 1875. Mendeleev even correctly predicted how the element would be discovered—with a spectroscope.

	Mendeleev's Predicted Properties for Gallium	*Actual Properties of Gallium*
Atomic mass	68	69.7
Density	5.9	5.903
Melting point	low	29.8°C
Formula of oxide	Ga_2O_3	Ga_2O_3

Figure 4-3 Comparison of predicted and experimentally determined properties of gallium.

Mendeleev also argued for liberal, social, and political reforms in Russia and was forced to leave the University of Saint Petersburg. Element 101, mendelevium, was named to honor his contribution to chemistry.

Mendeleev's table contained a few elements that appeared to be out of order when arranged by atomic masses so Mendeleev assumed the atomic masses were incorrect and arranged several elements by their properties rather than their atomic masses. Other elements discovered later—the lanthanide series—had no places on Mendeleev's table.

Englishman Henry Gwyn-Jeffreys Moseley discovered the concept of atomic number in 1912. This discovery allowed him to make a new periodic table, one arranged in order of increasing atomic numbers. His new table cleared up the remaining shortcomings of Mendeleev's table. Moseley's table left gaps for seven missing elements—these were later discovered. His experimental work was a remarkable achievement. In 1915, Mosely died at age 28 while fighting in World War I.

MODERN PERIODIC LAW

The modern **Periodic Law** states that the properties of the elements repeat themselves periodically if the elements are arranged in order of increasing atomic numbers. Many different versions of a periodic table have been developed on the basis of this law. The most common periodic table, shown in Figure 4-4, arranges similar elements in vertical columns called **groups** or **families**. The horizontal rows are called **periods**. The location of an element in the table can be used to predict its physical and chemical properties. The discovery of the Periodic Law is one of the most useful achievements in science. The Law is an extremely powerful tool to chemists, one that all chemists need to master.

Each chemical family is identified by a number of reference. Unfortunately, two rival numbering systems have existed for many years. In the United States, elements in the scandium group were labeled in Group IIIB and elements in the boron group were labeled IIIA. Elsewhere, most chemists followed the "official" International Union of Pure and Applied Chemistry (IUPAC) system which reversed the IIIA and IIIB designations. In an attempt to resolve this confusion, new group numbers were proposed that put scandium in Group 3 and boron in Group 13. Most chemists are not concerned with what the groups are called so long as everyone is consistent. U.S. chemical educators, however, prefer that boron be in Group IIIA. It is easier to explain to students that the representative elements are the "A" groups, and that for any **representative element**, the group number equals the number of valence electrons. Thus, boron, aluminum, and the rest of Group IIIA all have three valence electrons.

IA																	VIIIA
1	IIA											IIIA	IVA	VA	VIA	VIIA	2
3	4											5	6	7	8	9	10
11	12	IIIB	IVB	VB	VIB	VIIB	VIIIB	VIIIB	VIIIB	IB	IIB	13	14	15	16	17	18
19	20	21	22	23	24	25	26	27	28	29	30	31	32	33	34	35	36
37	38	39	40	41	42	43	44	45	46	47	48	49	50	51	52	53	54
55	56	57	72	73	74	75	76	77	78	79	80	81	82	83	84	85	86
87	88	89	104	105	106	107	108	109	110	111	112		114				

58	59	60	61	62	63	64	65	66	67	68	69	70	71
90	91	92	93	94	95	96	97	98	99	100	101	102	103

Figure 4-4 The most common form of the periodic table of the elements.

Ninety of the elements occur naturally on earth, some in extremely minute quantities. Technetium (atomic number 43), promethium (61), and the transuranium elements (atomic numbers greater than 92) are man-made. At the time of this writing, elements through atomic number 111 are claimed to have been synthesized. Several more new elements are likely to be made in the next year or two because of recent technological advances. With each new element scientists learn more about matter and the atom. Ultimately, this work is concerned with how the world was made. Initially no practical uses of these new elements are foreseen. But significant uses are possible someday, just as important uses were eventually found for man-made elements such as technetium, plutonium, americium, and californium.

Elements above atomic number 111 have temporary names and symbols, to be used until the existence and the discovery of the element can be verified and the official name determined by IUPAC.

One shortcoming of the periodic table of Figure 4-4 is that it does not show the true position of the **lanthanide series** or rare earth elements (elements 58-71) and the **actinide series** (elements 90-103) of elements. The long form of the periodic table, Figure 4-5, shows these elements in their correct positions.

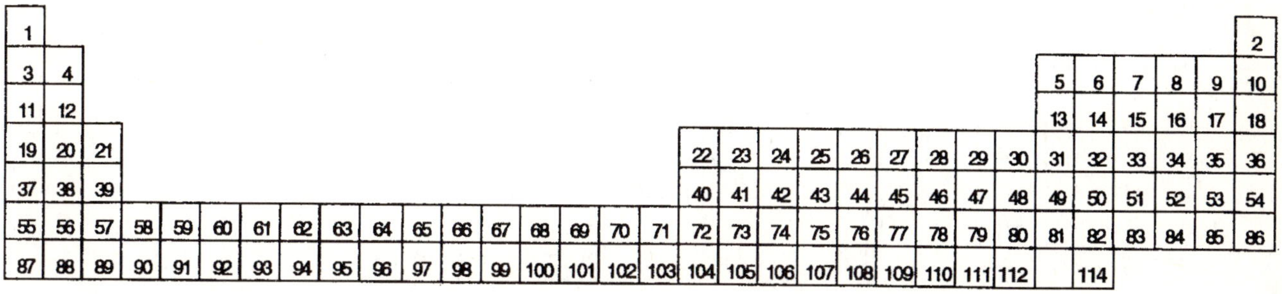

Figure 4-5 The long form of the periodic table of the elements.

In any form, the periodic table and the Periodic Law are useful in predicting such things as

- The properties of elements, such as size of atoms and electronegativity.
- The properties of compounds formed from the elements.
- The formulas of compounds.

PROPERTIES OF THE ELEMENTS

Atomic Radius

An atom can be thought of as a tiny sphere with the nucleus in the center. When two atoms are near each other, the electrons from atom 1 are attracted to the protons of atom 2, and the protons from atom 1 are repelled by the protons from atom 2. The two atoms find a balance between this attraction and repulsion and end up a certain distance apart. When two like atoms are bonded to each other, the **atomic radius** is half the distance between the two nuclei. Figure 4-6 shows the atomic radii of the elements; the trends in atomic radius are more apparent when displayed graphically in Figure 4-7. The units of the radii are Ångstroms, a non-SI metric unit. One Å equals 10–10 meter. . When the elements are arranged in order of atomic number, the atomic radius shows a trend that repeats itself. This trend is evident only when the elements are arranged in order of atomic number; arranging the elements in alphabetical order would not reveal the trend. The trend can be used to predict the radius of undiscovered elements.

Electronegativity

All atoms (except the noble gases) have fewer than eight valence electrons. **Valence electrons** are the electrons in the outermost energy level. For the representative elements, the number of valence electrons is equal to the Group Number; for example carbon is in Group IVA and therefore has four valence electrons. The nucleus of all atoms (except the noble gases) has a tendency to attract electrons from outside the atom. **Electronegativity** is the power of an atom to attract electrons towards itself when it is part of a compound. Electronegativity values are numbers from a scale originally set up by Nobel Laureate Linus Pauling and later slightly modified by others. Figure 4-8 shows trends in electronegativities among the elements; these trends are more apparent when graphically displayed in Figure 4-9.

H																	He
0.32																	0.93
Li	Be											B	C	N	O	F	Ne
1.23	0.90											0.82	0.77	0.75	0.73	0.72	0.71
Na	Mg											Al	Si	P	S	Cl	Ar
1.54	1.36											1.18	1.11	1.06	1.02	0.99	0.98
K	Ca	Sc	Ti	V	Cr	Mn	Fe	Co	Ni	Cu	Zn	Ga	Ge	As	Se	Br	Kr
2.03	1.74	1.44	1.32	1.22	1.18	1.17	1.17	1.16	1.15	1.17	1.25	1.26	1.22	1.20	1.16	1.14	1.89
Rb	Sr	Y	Zr	Nb	Mo	Tc	Ru	Rh	Pd	Ag	Cd	Sn	Sn	Sb	Te	I	Xe
2.16	1.91	1.62	1.45	1.34	1.30	1.27	1.25	1.25	1.28	1.34	1.41	1.44	1.41	1.40	1.36	1.33	1.31
Cs	Ba	La	Hf	Ta	W	Re	Os	Ir	Pt	Au	Hg	Tl	Pb	Bi	Po	At	Rn
2.35	1.98	1.25	1.44	1.34	1.30	1.28	1.26	1.27	1.30	1.34	1.49	1.48	1.47	1.46	1.53	1.47	-
Fr	Ra	Ac															
-	-	-															

Ce	Pr	Nd	Pm	Sm	Eu	Gd	Tb	Dy	Ho	Er	Tm	Yb	Lu
1.65	1.65	1.64	1.63	1.62	1.85	1.61	1.59	1.59	1.58	1.57	1.56	1.70	1.56
Th	Pa	U	Np	Pu	Am	Cm	Bk	Cf	Es	Fm	Md	No	Lr
1.65	-	1.42	-	1.08	-	-	-	-	-	-	-	-	-

Figure 4-6 Atomic radii of the elements in units of Ångstroms.

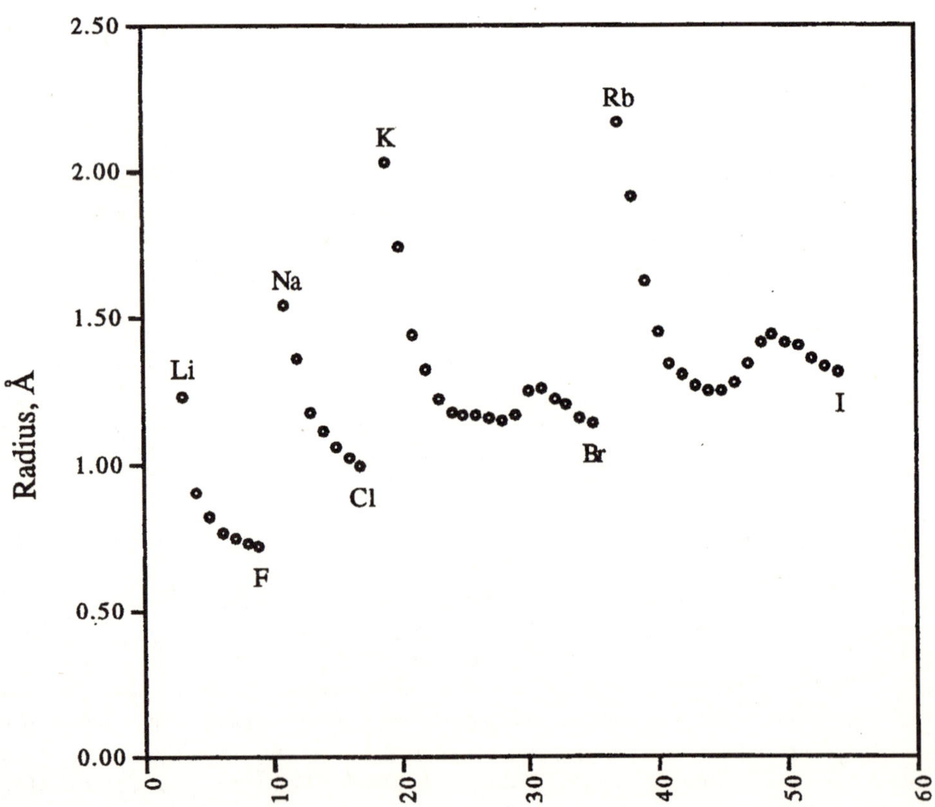

Figure 4-7 Trends in atomic radius are more obvious when graphed. In order to make the trends more apparent, hydrogen and the noble gases are not included.

H 2.1																	He
Li 1.0	Be 1.6											B 2.0	C 2.5	N 3.0	O 3.4	F 4.0	Ne
Na 0.9	Mg 1.3											Al 1.6	Si 1.9	P 2.2	S 2.6	Cl 3.2	Ar
K 0.8	Ca 1.0	Sc 1.4	Ti 1.5	V 1.6	Cr 1.7	Mn 1.6	Fe 1.8	Co 1.9	Ni 1.9	Cu 1.9	Zn 1.7	Ga 1.8	Ge 2.0	As 2.2	Se 2.6	Br 3.0	Kr
Rb 0.8	Sr 1.0	Y 1.2	Zr 1.3	Nb 1.6	Mo 2.2	Tc 1.9	Ru 2.2	Rh 2.3	Pd 2.2	Ag 1.9	Cd 1.7	Sn 1.8	Sn 2.0	Sb 2.1	Te 2.1	I 2.7	Xe
Cs 0.8	Ba 0.9	La 1.1	Hf 1.3	Ta 1.5	W 2.4	Re 1.9	Os 2.2	Ir 2.2	Pt 2.3	Au 2.6	Hg 2.0	Tl 2.0	Pb 2.3	Bi 2.0	Po 2.0	At 2.2	Rn
Fr 0.7	Ra 0.9	Ac 1.1															

Ce 1.1	Pr 1.1	Nd 1.1	Pm 1.1	Sm 1.2	Eu 1.2	Gd 1.2	Tb 1.1	Dy 1.2	Ho 1.2	Er 1.2	Tm 1.3	Yb 1.1	Lu 1.3
Th 1.3	Pa 1.5	U 1.4	Np 1.4	Pu 1.3	Am 1.3	Cm 1.3	Bk 1.3	Cf 1.3	Es 1.3	Fm 1.3	Md 1.3	No 1.3	Lr 1.3

Figure 4-8 Electronegativities of the elements.

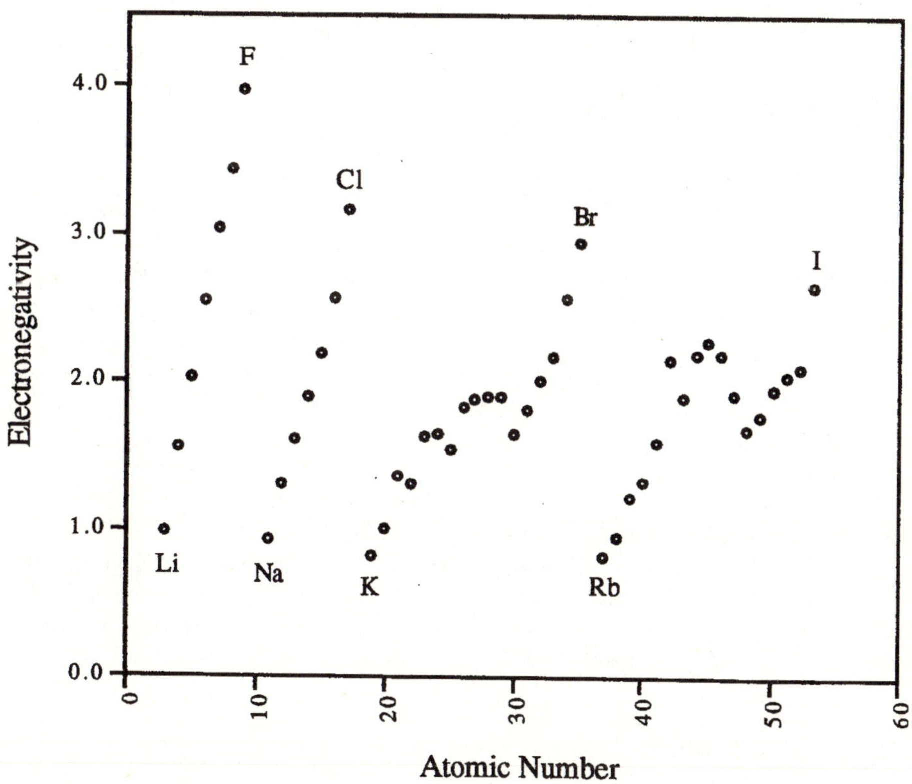

Figure 4-9 Trends in electronegativities of the elements. In order to make the trends more apparent, hydrogen and the noble gases are not included.

PROPERTIES OF COMPOUNDS

Chemical compounds generally fall into one of two broad categories: ionic compounds and covalent compounds.

Ionic compounds are always solid at room temperature. They have high melting points, over 350°C and often as high as 1000°C. Most ionic compounds dissolve in water; few dissolve in nonpolar solvents like paint thinner. Ionic compounds always conduct an electric current when dissolved in water or when heated to the liquid state. They do not burn and few have odors.

Covalent compounds are also called molecular compounds. Although some are solid at room temperature, many are liquids or gases. Their melting points are low—less than 350°C. A few covalent compounds dissolve in water; many dissolve in nonpolar solvents. Few conduct an electric current. Most will burn and many have pronounced odors.

It is possible to predict whether a compound will have properties of an ionic or a covalent compound. The most accurate prediction requires calculating the difference in electronegativity between two atoms bonded to each other. When the difference in electronegativity values is large (1.7 or greater), the major attraction between the atoms is an ionic bond. When the difference in electronegativity values is small (less than 1.7), the major attraction between the atoms is a covalent bond. Most bonds have a mixture of ionic and covalent character but one character predominates and it determines the properties of the compound.

A faster but slightly less accurate prediction can be made by first classifying the elements as metals or nonmetals. Metals, as seen in Figure 4-8, have low electronegativities; nonmetals have high electronegativities. A bond between a metal and a nonmetal will therefore generally have a large difference in electronegativities and will be an ionic bond. A bond between a nonmetal and a nonmetal will generally have a low difference in electronegativities and will be a covalent bond.

FORMULAS OF COMPOUNDS

The **Octet Rule** states that atoms tend to gain, lose, or share electrons in order to end up with the same electron configuration of one of the noble gases. All the noble gases have eight valence electrons—except for helium, which has two valence electrons. Thus, atoms tend to gain, lose, or share electrons in order to end up with eight valence electrons. Hydrogen is the exception to the rule of eight; this nonmetal is usually placed above the alkali metals on the periodic table because it, too, has one valence electron. Unlike the alkali metals, hydrogen atoms tend to gain, lose, or share electrons to end up with two valence electrons—just as the noble gas helium has two valence electrons. So, while other elements have a rule of eight, hydrogen has a rule of two valence electrons.

Metal atoms can satisfy the Octet Rule by losing one or more electrons to become positively charged ions or **cations**. It is easy to predict the charge of the ions of the alkali metals, the alkaline earth metals, aluminum, and a few other metals that always form the same ion. The group number of the representative elements is the number of valence electrons; metals will lose all their valence electrons. The group number of representative metals is the positive charge of the ion.

Group	*Charge of Ion*	*Example*	
IA	1+	Na^{1+}	sodium ion
IIA	2+	Ca^{2+}	calcium ion
IIIA	3+	Al^{3+}	aluminum ion

Nonmetals atoms can satisfy the Octet Rule in two ways: (1) by sharing electrons with another nonmetal to form a covalent bond, or (2) by gaining one or more electrons to form negatively charged ions or **anions**. It is easy to predict the charge of nonmental anions. The group number is the number of valence electrons; subtract the group number from 8 to find the negative charge of the nonmetal monoatomic ion. Names of nonmetal monoatomic ions always end with an **-ide** suffix.

Group	*Charge of Ion*	*Example*	
VA	3–	N^{3-}	nitride ion
VIA	2–	O^{2-}	oxide ion
VIIA	1–	Cl^{1-}	chloride ion

Binary compounds contain only two elements and their formulas are easily predicted using the periodic table. Consider the compound calcium chloride which is sometimes used to melt ice on sidewalks. Calcium is a metal in group IIA, therefore calcium has two valence electrons. By applying the Octet Rule, we see calcium must lose its two valence electrons and so calcium has a 2+ charge, Ca^{2+}. The nonmetal chlorine is in group VIIA and has seven valence electrons. Chlorine atom needs to gain one electron to satisfy the Octet Rule and so chloride ion has a charge of 1–, Cl^{1-}.

Once you've used the periodic table to predict the charge of the metal cation and the nonmetal anion, list the positive ion first and the negative ion second. Then simply "criss-cross" the charges to determine the subscripts of the compound's formula. The charge of calcium (2) becomes the subscript of chlorine. The charge of chlorine (1) becomes the subscript of calcium. Subscripts of 1 are not written; when an atom does not have a subscript it means there is one atom of that element. So instead of Ca_1Cl_2, calcium chloride's formula is $CaCl_2$.

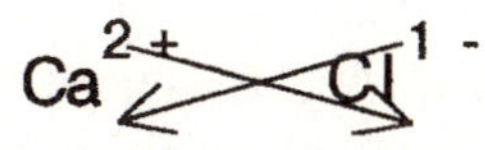

This "criss-cross" method works in all cases except where the subscripts are equal. For example, criss-crossing the charges of calcium ion (Ca^{2+}) and oxide ion (O^{2-}) gives the incorrect formula Ca_2O_2. Formulas of ionic compounds are always written as simplest whole-number ratios. For calcium oxide, the simplest ratio is not 2 to 2, but 1 to 1. The correct formula is CaO.

The names of binary ionic compounds list the metal name first, followed by the nonmetal name with an -ide suffix.

Procedure

I. STRUCTURE OF THE PERIODIC TABLE

Use the blank periodic table in Figure 4-10 to complete the following steps. Use colored pencils to help identify the selected feature. You may use your textbook as a reference.

1. Number the periods and the groups.

2. Label the following chemical families (groups):
   a. Alkali metals
   b. Alkaline earth metals
   c. Halogens
   d. Noble gases

3. Sketch a dark black line separating the metals from the nonmetals.

4. Label the following elements:
   a. Representative elements
   b. Transition elements
   c. Lanthanide series (formerly called rare earth elements)
   d. Actinide series

II. PROPERTIES OF ELEMENTS

A. Atomic Radius

1. Period 1 is not included in Figure 4-7. Find the elements in Period 2 in Figure 4-7 and note the trend in atomic radius as atomic numbers increase in Period 2. Find the elements in Period 3 and note the trend in Period 3. What generalization can you draw about the trend in radius as atomic numbers increase in a period?

2. Find the alkali metals in Figure 4-7 and note the trend in atomic radius as atomic numbers increase in this family. Find the halogens and note the trend within that family. What generalization can you draw about the trend in atomic radius as atomic numbers increase in a family?

B. Electronegativity

1. Period 1 is not included in Figure 4-9. Find the elements in Period 2 and note the trend in electronegativity. Note the trend in Period 3. What generalization can you make about the trend in electronegativity as atomic numbers increase in a period?

2. Find the alkali metals in Figure 4-7 and note the trend in atomic radius as atomic numbers increase in this family. Find the halogens and note the trend within that family. What generalization can you draw about the trend in atomic radius as atomic numbers increase in a family?

C. Charges of Monoatomic Ions

1. Write the symbol (including charge) in the appropriate position in the blank periodic table of Figure 4-11 of the following representative element metal ions:
   a. Lithium ion
   b. Sodium ion
   c. Potassium ion
   d. Magnesium ion
   e. Calcium ion
   f. Strontium ion
   g. Barium ion
   h. Aluminum ion

2. Write the symbol (including charge) in the appropriate position in Figure 4-11 of the following monoatomic nonmetal ions:
   a. Fluoride ion
   b. Chloride ion
   c. Bromide ion
   d. Iodide ion
   e. Oxide ion
   f. Sulfidc ion
   g. Nitride ion

D. Predictions for an Undiscovered Element

1. a. Draw a new element box on Figure 4-11 for the undiscovered element with atomic number 117.
   b. Choose a name and symbol for element 117 and write the symbol on Figure 4-11.
   c. What name is given to the family to which element 117 belongs?
   d. How many valence electrons would it have?
   e. Would element 117 have properties of a metal, nonmetal, or metalloid?

2. a. Use Figure 4-12 to graph the atomic radii of five elements in the same family as element 117.
   b. Use the graph to predict the atomic radius of element 117.

3. a. Use Figure 4-13 to graph the electronegativity of five elements in the same family as element 117.
   b. Use the graph to predict the electronegativity of element 117.

4. Which of the following atomic masses would element 117 have, based on Döbereiner's Triad method? Show calculations that verify your choice.
   a. 283
   b. 293
   c. 303
   d. 313

III. FORMULAS OF BINARY IONIC COMPOUNDS

Refer to Figure 4-14. Use each element's position on the periodic table to predict the charge of the element's monoatomic ion. Use the charges of the ions to predict the formula of each compound.

Name ________________________ Section ____________ Date ____________

Prelab Questions for Experiment 4

1. Why is the periodic table format in Figure 4-4 more frequently used than the long form of the periodic table in Figure 4-5?

   ____

   ____

2. What is the principal difference between the arrangement of elements on Mendeleev's periodic table and our present periodic table (as shown in Figure 4-4)?

   ____

   ____

3. Make a list of three sequences of elements on the present periodic table that would appear to be out of place if arranged in order of atomic masses.

   1. ____________ ____________ ____________

   2. ____________ ____________ ____________

   3. ____________ ____________ ____________

4. Complete the following table by comparing the properties of ionic and covalent compounds.

	Ionic Compounds	*Covalent Compounds*
Melting Point		
Solubility in Water		
Conduction of Electric Current		
Flammability		
Odor		

5. Naphthalene has a melting point of 80.5°C, is soluble in nonpolar solvents but not water, and does not conduct an electric current. Is naphthalene likely an ionic compound or a covalent compound?

____

6. List the elements in Period 2 of the periodic table.

____

____

7. List the elements in the alkali metal family.

____

____

8. Find the electronegativity (EN) of each atom in the following compounds. Find the difference in electronegativity (ΔEN) between the atoms in each pair. Predict whether the bond between the two atoms is ionic or covalent.

	EN1 (atom 1)	*EN2 (atom 2)*	*$\Delta EN = EN2 - EN1$*	*Bond Type*
K-F				
Li-Br				
I-Cl				
Cl-Cl				

9. a. What numerical difference in electronegativity results in a covalent bond?

____

b. What numerical difference in electronegativity results in an ionic bond?

____

10. a. In general, what type of bond is formed when a metal bonds with a nonmetal?

____

b. In general, what type of bond is formed when a nonmetal bonds with a nonmetal?

____

Name ______________________________ Section ____________ Date ______________

Report for Experiment 4

I. STRUCTURE OF THE PERIODIC TABLE

Follow steps 1-4 in Procedure I to fill in Figure 4-10.

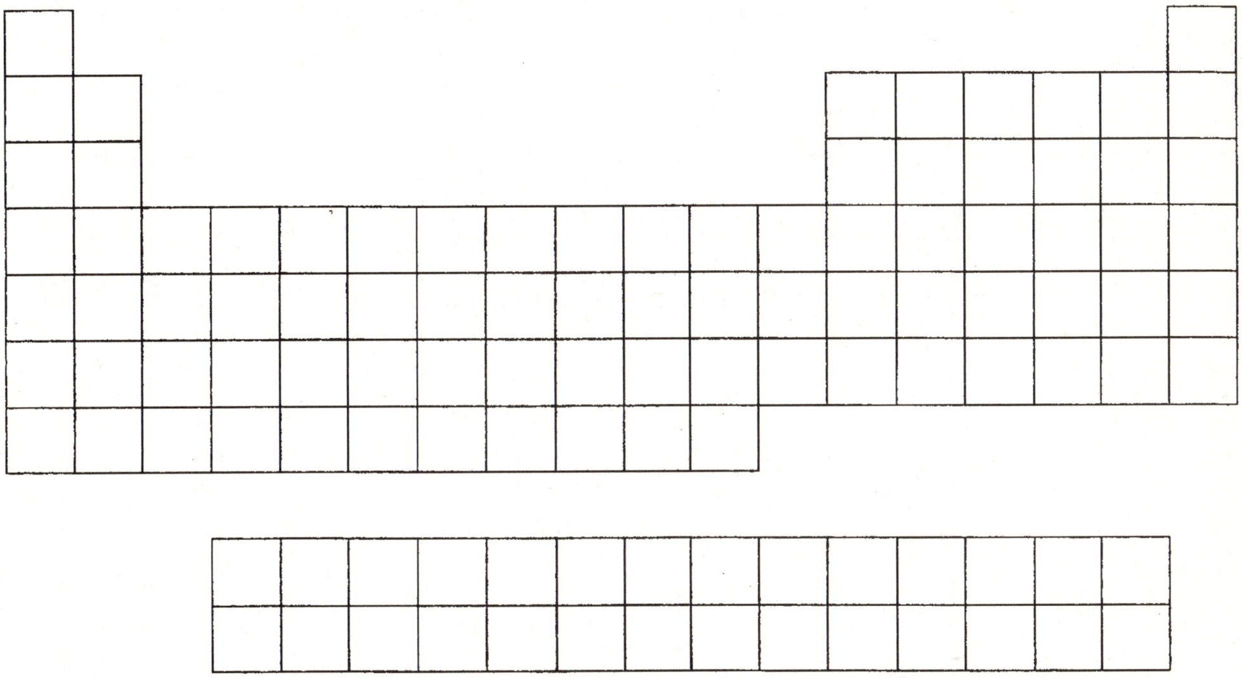

Figure 4-10

II. PROPERTIES OF ELEMENTS

A. Atomic Radius

1. What generalization can you make about the trend in atomic radius as atomic number increases in a period?

____

____

2. What generalization can you make about the trend in atomic radius as atomic number increases in a group (family)?

____

____

B. Electronegativity

1. What generalization can you make about the trend in electronegativity as atomic number increases in a period?

____

____

2. What generalization can you make about the trend in electronegativity as atomic number increases in a group (family)?

____

____

C. Charges of Monoatomic Ions

1. Follow Steps 1 and 2 in Procedure II C to fill in Figure 4-11.

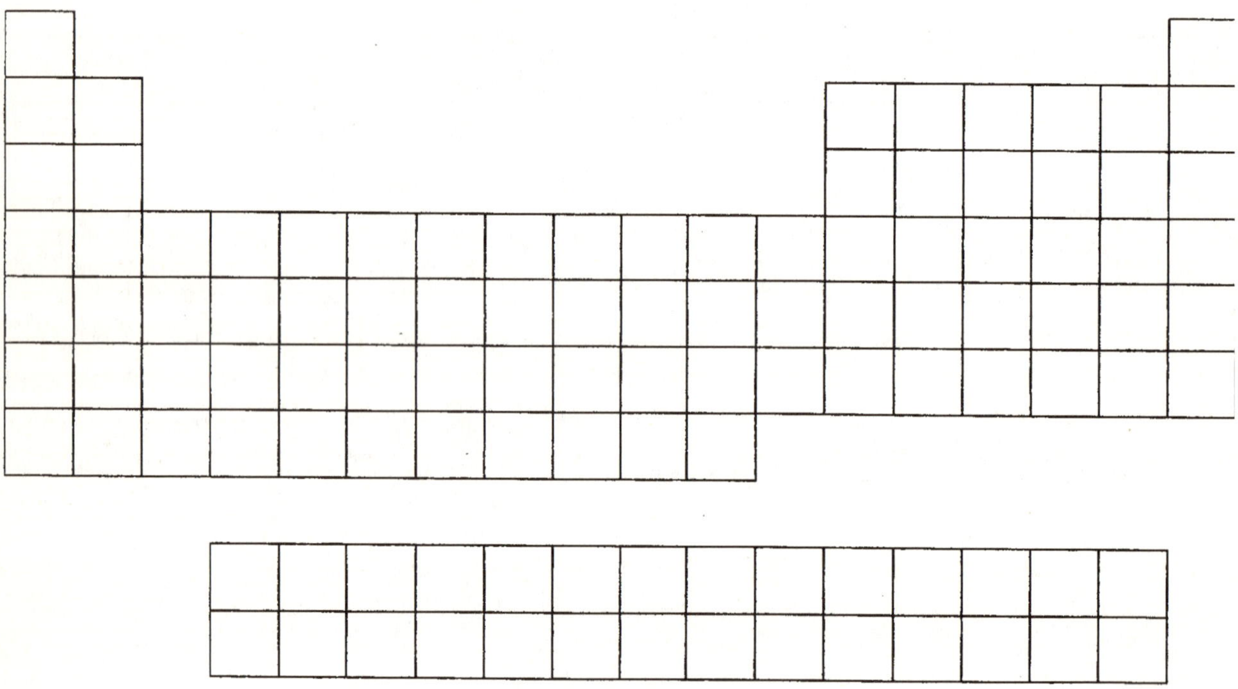

Figure 4-11

2. Explain how a representative element's position on the periodic table is helpful in determining the charge of its monoatomic ion.

____

____

Name ____________________ Section ____________ Date ____________

D. Predictions for an Undiscovered Element

1. a. Draw element 117 and its symbol on Figure 4-11.

   b. Proposed name of element 117 ____________________

   Proposed symbol of element 117 ____________________

   c. Family (group) of element 117 ____________________

   d. Number of valence electrons for element 117 ____________________

   e. Classify element 117 as metal, nonmetal, or metalloid ____________________

2. a. Graph the atomic radii of elements in the same family as element 117 in Figure 4-12.

   b. Predicted atomic radius of element 117 ____________________

3. a. Graph the electronegativity of elements in the same family as element 117 in Figure 4-13.

   b. Predicted electronegativity of element 117 ____________________

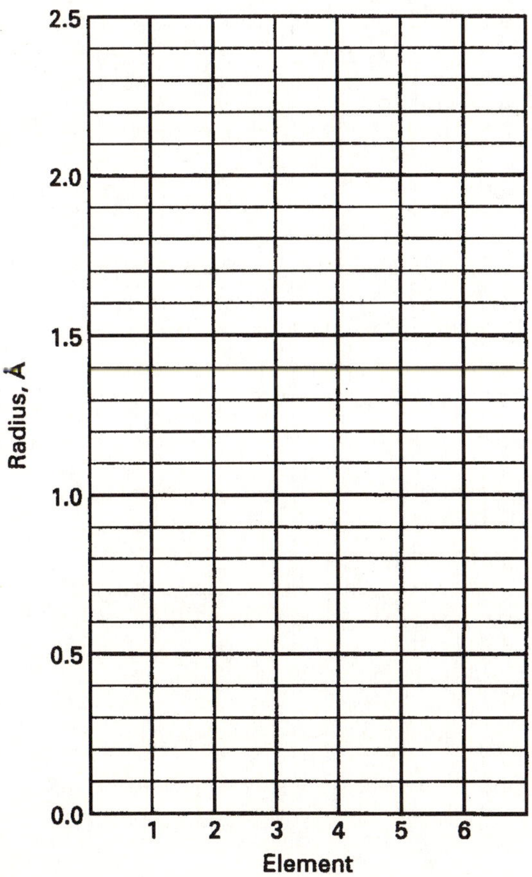

Figure 4-12

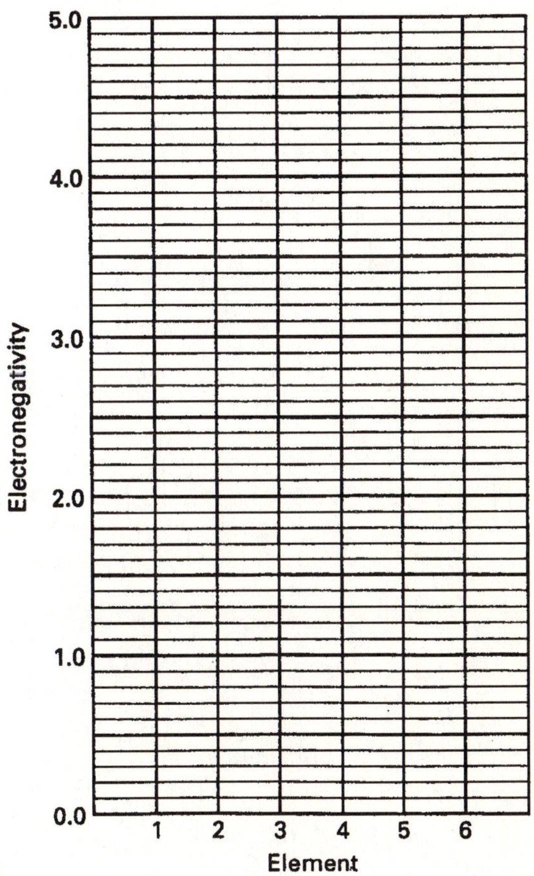

Figure 4-13

4. Predicted atomic mass of element 117

   Calculations:

III. FORMULAS OF BINARY IONIC COMPOUNDS

	Name of compound	*Formula (with charge) of cation*	*Formula (with charge) of anion*	*Formula of compound*
a.	sodium bromide			
b.	calcium iodide			
c.	barium oxide			
d.	aluminum chloride			
e.	lithium nitride			
f.	magnesium sulfide			
g.	potassium nitride			
h.	strontium fluoride			
i.	aluminum oxide			
j.	potassium sulfide			
k.	sodium iodide			

Figure 4-14

Name ______________________ Section ____________ Date ______________

Related Questions for Experiment 4

1. Classify each compound as ionic or covalent by using the metal/nonmetal technique—if there is a metal, the compound is ionic; covalent compounds do not contain metals. If an ionic compound, write the formula and charge of each ion (some ions are monoatomic and some are polyatomic), then use the charges of the ions to predict the formula of the ionic compound. If a covalent compound, do not write the formulas and charges of ions; instead, place a large X in the boxes for cations and anions. Remember, covalent compounds do not contain ions! If a covalent compound, use the prefixes in the compound's name to write the formula.

	Name of compound	*Ionic compound or covalent compound?*	*Formula and charge of cation (if ionic compound)*	*Formula and charge of anion (if ionic compound)*	*Formula of compound*
a.	sodium sulfide				
b.	sodium sulfate				
c.	sulfur dioxide				
d.	calcium hydroxide				
e.	potassium nitrate				
f.	barium chloride				
g.	phosphorus trichloride				
h.	sodium phosphate				
i.	calcium carbonate				
j.	calcium bicarbonate				
k.	nitrogen dioxide				
l.	potassium oxide				
m.	carbon monoxide				

2. Classify each compound as ionic or covalent—use the metal/nonmetal technique. Write the name of each compound. Remember that covalent compounds use prefixes as part of their name, ionic compounds do not use prefixes.

	Formula	*Ionic compound or covalent compound?*	*Name*
a.	NaI		
b.	$Mg(NO_3)_2$		
c.	CO_2		
d.	K_2S		
e.	K_2SO_4		
f.	SO_3		
g.	Na_2CO_3		
h.	$NaHCO_3$		
i.	P_2O_5		
j.	$Mg_3(PO_4)_2$		
k.	BaO		
l.	$Ba(OH)_2$		
m.	LiBr		
n.	CS_2		
o.	CaS		
p.	$CaSO_4$		

Experiment 5
Hydrocarbons

Reference: *General, Organic and Biological Chemistry: An Integrated Approach, Chapter 4.*

Purpose: Using molecular models, you will investigate the relationship between a structural formula and a three-dimensional molecule. The models will be used to illustrate constitutional and geometric isomerism. You will perform chemical tests that illustrate important chemical properties of hydrocarbons.

Introduction

Hydrocarbons are organic compounds containing only hydrogen and carbon. Their major natural sources are petroleum, natural gas, and coal tar. On the basis of chemical properties and structures, hydrocarbons are subdivided into two broad classes: **aliphatic** and **aromatic**. Aliphatic hydrocarbons can be further subdivided into alkanes, alkenes, alkynes, cycloalkanes, and cycloalkenes.

ALIPHATIC HYDROCARBONS

Alkanes

Alkanes are hydrocarbons in which the carbon atoms in the molecule are connected only by a single covalent bond. Some typical alkanes are shown below:

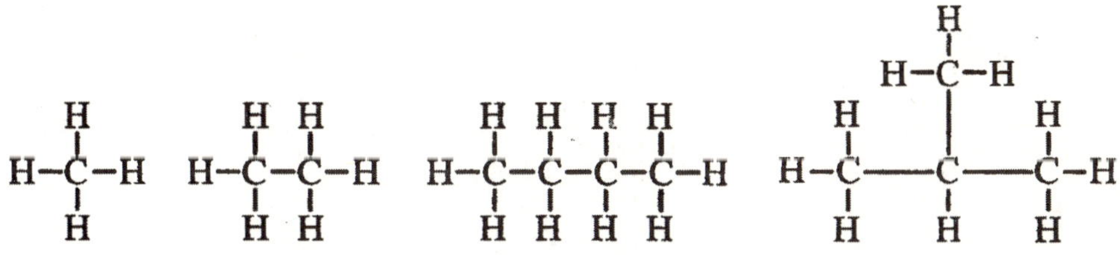

methane ethane butane 2-methylpropane

The two four-carbon compounds, butane and 2-methylpropane, have the same molecular formula, C_4H_{10}, but different molecular structures. They are said to be **isomers** and the kind of isomerism they exhibit is called structural or **constitutional isomerism**.

As the number of carbon atoms in alkanes increases, the number of constitutional isomers will also increase. Three constitutional isomers of C_5H_{12} are possible and C_6H_{14} has five isomers. The ability of carbon atoms to bond together one after another to form long chains in different patterns accounts for the tremendous number of carbon compounds.

Alkenes

Hydrocarbons containing double covalent bonds between carbon atoms are known as **alkenes**. Typical alkenes are:

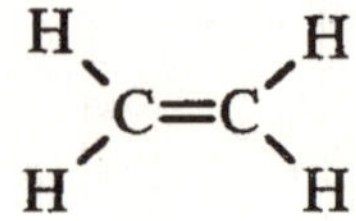

ethene

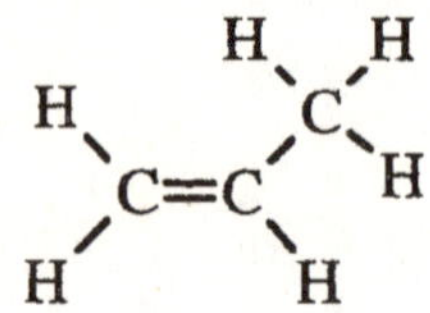

propene

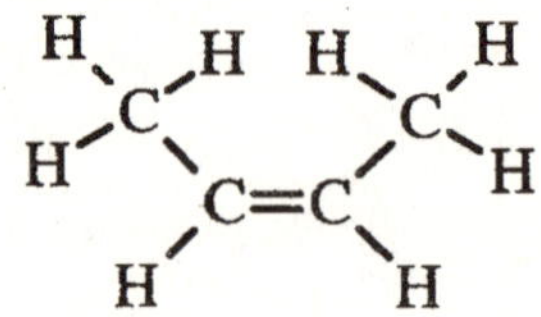

cis-2-butene

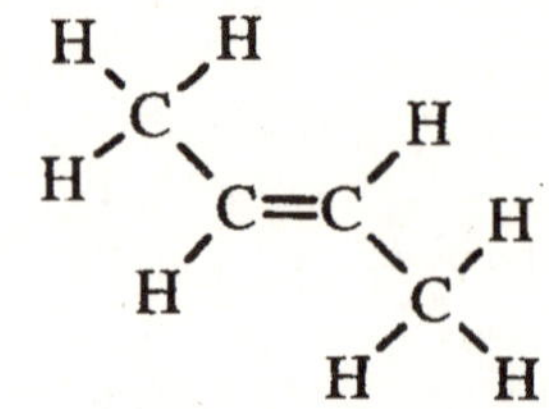

trans-2-butene

The existence of a carbon-to-carbon double bond gives rise to another type of isomerism: **geometric (or *cis-trans*) isomerism**. A single bond between carbon atoms, such as found in alkanes, does not restrict the rotation of atoms around that bond; the carbon atoms can twist freely. However, a double bond between two carbon atoms is structurally rigid, preventing free rotation of the carbon atoms. When two specified atoms or groups of atoms attached to the double-bonded carbon atoms appear on the same side of the bond, the molecule is a *cis* isomer. A *trans* isomer is formed when two specified atoms or groups of atoms appear on opposite sides of the bond (see *cis*-2-butene and *trans*-2-butene structures). Geometric isomers are not limited to hydrocarbons; note the following halogenated compounds:

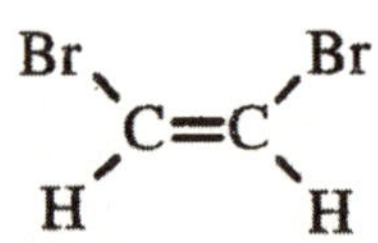

cis-1,2-dibromoethene

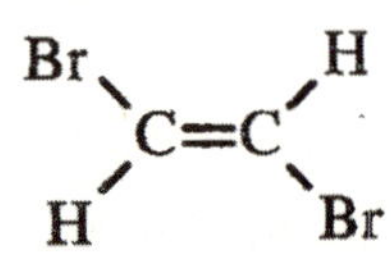

trans-1,2-dibromoethene

Alkynes

Alkynes are compounds having a triple bond between two carbons. They are not of biological importance and are not studied in this experiment.

Compounds with double or triple bonds are said to be **unsaturated hydrocarbons**; those with single bonds are **saturated hydrocarbons**.

Cycloalkanes and Cycloalkenes

Carbon atoms can form stable rings of various sizes, with the possibility of both single and double bonds. For example:

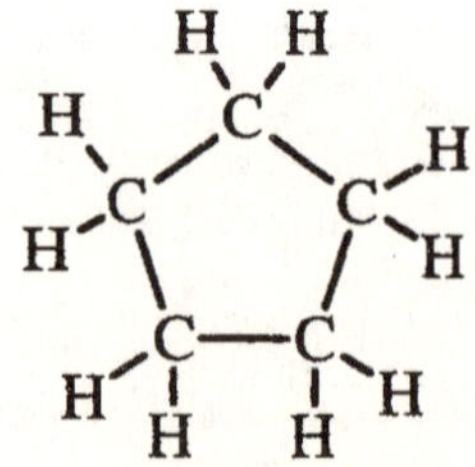

cyclopentane

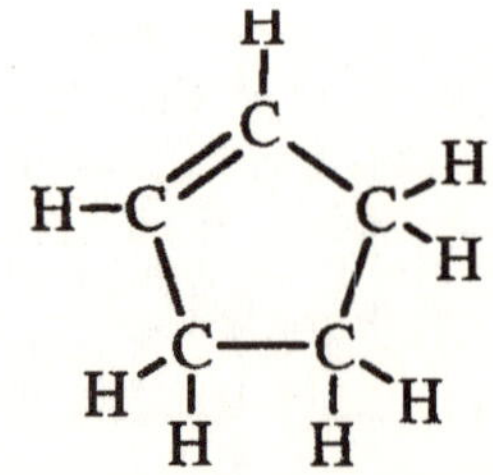

cyclopentene

In practice, chemists use a shorthand representation of the above ring structures:

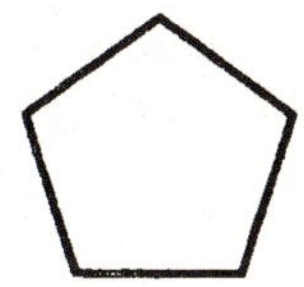

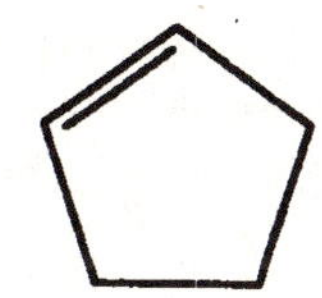

If elements other than carbon and hydrogen are involved in the compound, they will be separately identified. Here are some examples:

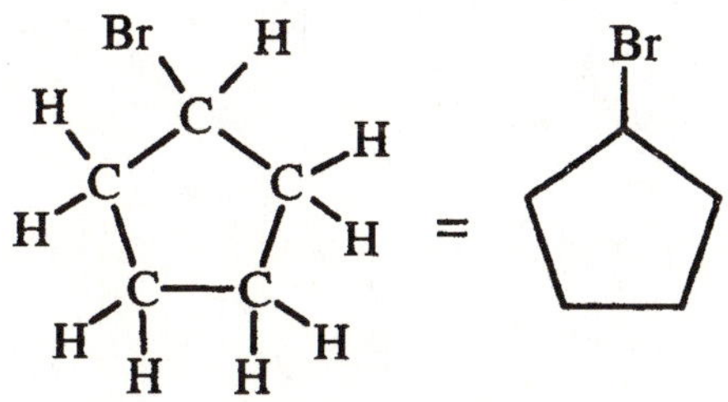

bromocyclopentane

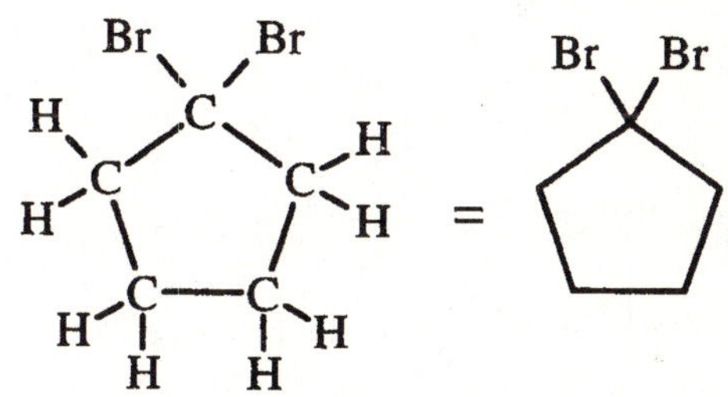

1,1-dibromocyclopentane

When carbon atoms are arranged in a ring, free rotation around even the carbon-to-carbon single bond is restricted. This leads to the existence of *cis* and *trans* isomers in ring compounds.

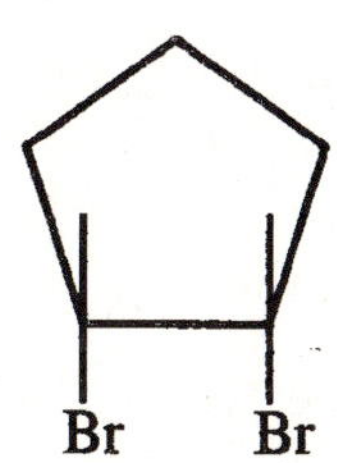

***cis*-1,2-dibromocyclopentane**

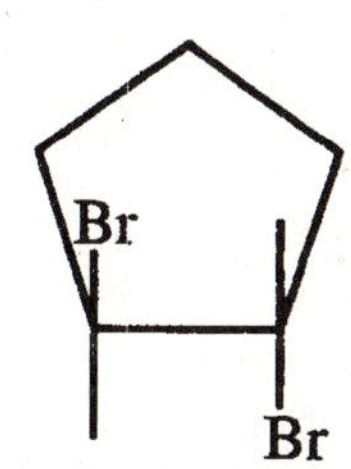

***trans*-1,2-dibromocyclopentane**

In this shorthand representation, the two bromine atoms are on the same side of the ring in the *cis* isomer and on opposite sides of the ring in the *trans* isomer. Each of the five carbon atoms in the ring has two bonds to other carbon atoms in the ring and two bonds to atoms not in the ring. After you make models of these structures, you will see that one of the later bonds is to an atom above the ring and the other is to an atom below the ring. In these shorthand structures, the atoms above the ring are attached to bonds pointing straight up, while bonds pointing straight down are to atoms below the ring.

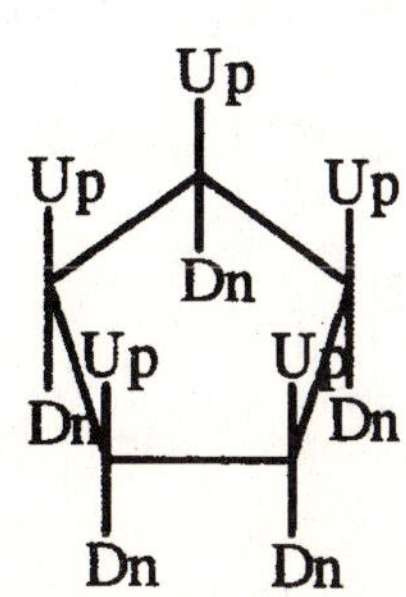

AROMATIC HYDROCARBONS

The most important of the cyclic hydrocarbons is benzene. Benzene and its derivatives compose the class of compounds called the **aromatic hydrocarbons**. Benzene has the molecular formula C_6H_6, but its structure cannot be truly represented by simple single and double bonds. The six carbon atoms in a molecule of benzene are in a ring and are symbolically represented as:

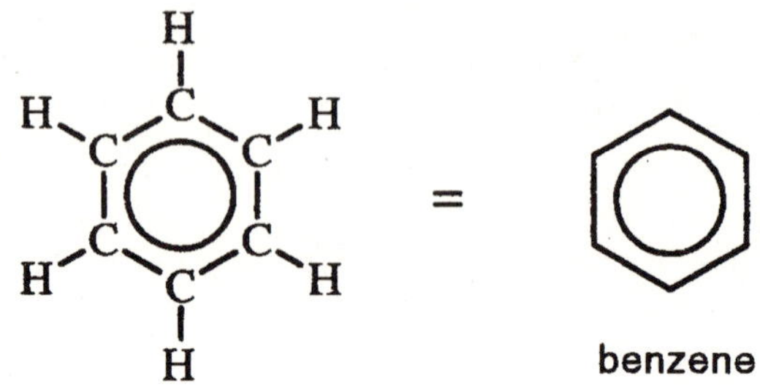

benzene

If atoms or groups replace hydrogen atoms of benzene, they will be separately identified. Thus:

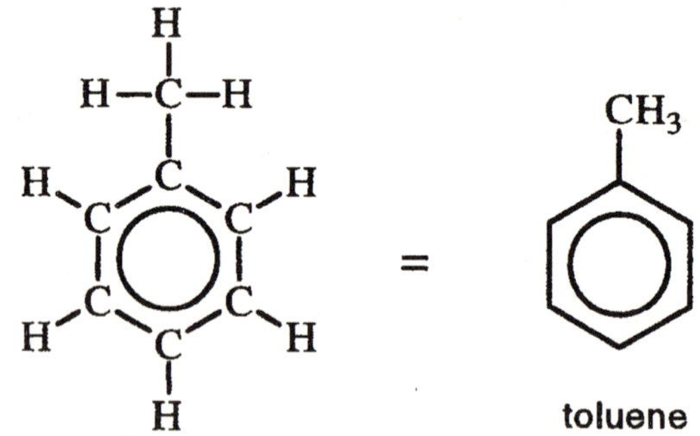

toluene

REACTIONS OF HYDROCARBONS

Oxidation

Hydrocarbons burn in an excess of oxygen to form carbon dioxide and water. Because the reaction is highly exothermic, hydrocarbons are often used as fuels. The combustion of methane in natural gas is an example:

$$CH_4 + 2\ O_2 \rightarrow CO_2 + 2\ H_2O + \text{Energy}$$

Limited supplies of oxygen will produce incomplete combustion, liberating massive quantities of carbon as smoke. If you used the Bunsen burner in Experiment 1, recall the smoky, luminous Bunsen burner flame when the burner air inlet was closed.

Reaction with Bromine

Bromine reacts differently at room temperature with each type of hydrocarbon studied in this experiment as illustrated by the following equations.

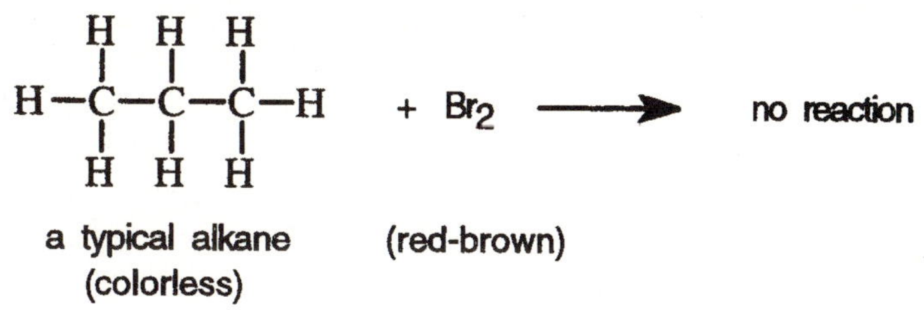

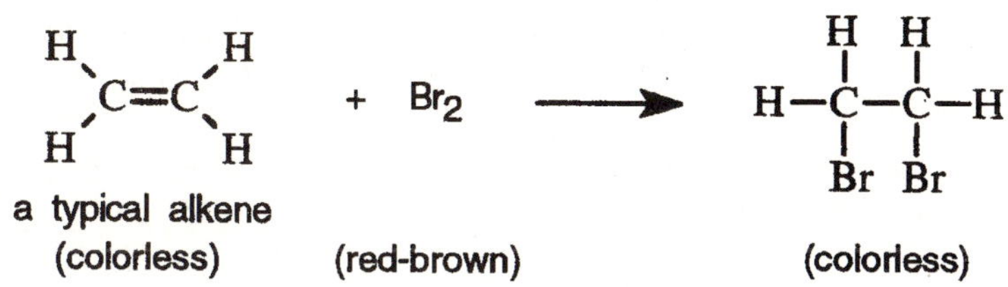

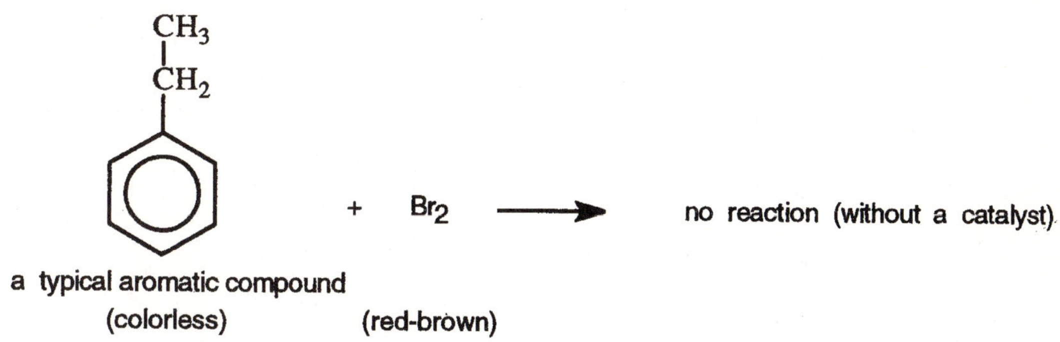

Whenever a reaction occurs, the red-brown color of bromine disappears and colorless products are formed. When there is no reaction, the red-brown color of bromine persists. Under the conditions specified in this experiment, treatment of the hydrocarbon with bromine provides a means for identifying alkenes.

MOLECULAR MODELS

The "ball-and-stick" molecular model set you will use in this experiment consists of colored wooden or plastic spheres, representing atoms, drilled to receive connecting bonds. Your instructor will furnish the color code for your model set. Carbon atom spheres have four drilled holes that represent the four covalent bonds that carbon atoms always form. Hydrogen atoms form one bond, bromine atoms form one bond, and oxygen atoms form two bonds; spheres for these atoms usually have the appropriate number of holes.

Single covalent carbon-to-carbon bonds are represented by wooden or plastic pegs that can be inserted into the holes of the atoms. Some model sets use shorter pegs to represent bonds between carbon and any other element. Double bonds are created either with two bent-pegs or with two bendable spring-pegs.

The structures you will assemble in this experiment closely approximate the geometry of the molecules they represent. Two structures are identical if they are superimposable—that is, if one structure can be placed "on top" of another so that all the colored spheres coincide.

Procedure

I. MOLECULAR MODELS: ISOMERISM

A. Methane

1. Construct two models of the methane (CH4) molecule. Superimpose one model on top of the other to show they are identical structures. Sketch the three-dimensional structure of the methane molecule.

2. Hold one of the models on the stage of an overhead projector so that the projection screen shows all five atoms. You must hold the model with only two hydrogen atoms resting on the stage as in Figure 5-1. Held properly, the shadows of the bonds (pegs) will be at right angles to each other. Draw the shadow on your Report Sheet. Alternatively, hold one of the models above a sheet of white paper so that the shadow of the model appears on the paper. It may be necessary to dim the room lights and to place a lamp above the model.

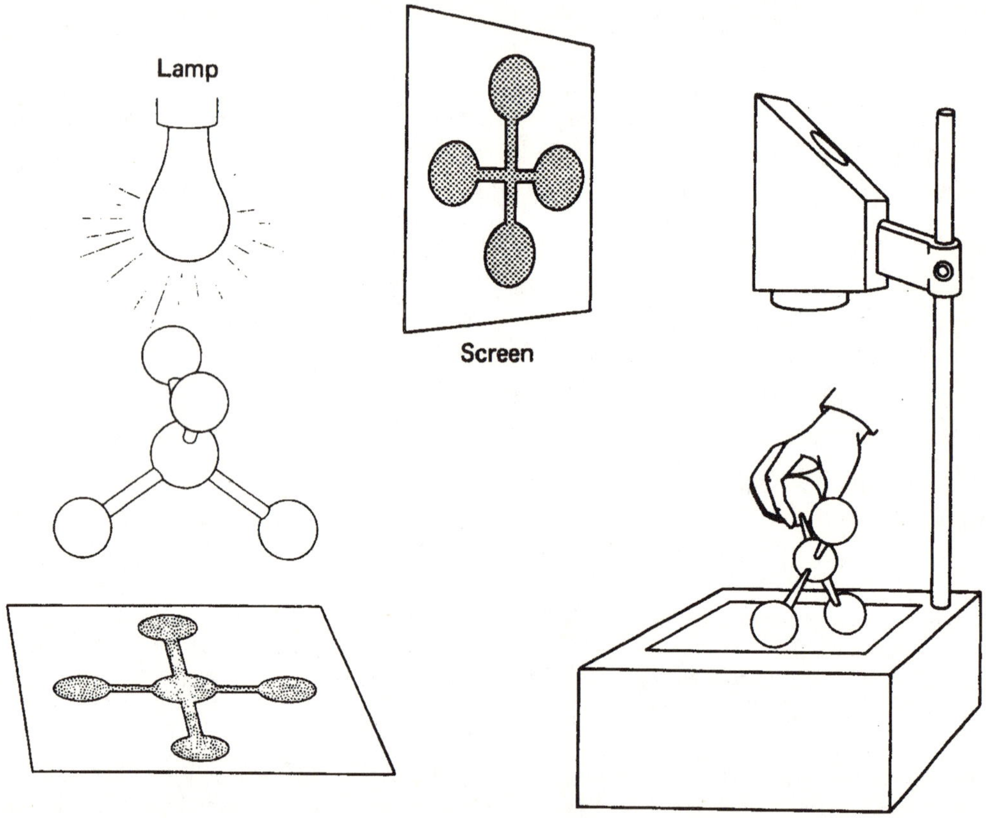

Figure 5-1 Two methods of projecting a shadow of a methane molecule.

3. Draw the structural formula of methane. Notice the similarities between the shadow and the structural formula; both have right angles between the bonds. Neither the shadow nor the structural formula have the 109.5° tetrahedral bond angles of the three-dimensional methane molecule.

4. Remove one of the hydrogen atoms from one of the structures and replace it with a bromine atom. Place the structure aside for reference.

5. On the second molecule of methane you constructed in Step 1, replace one hydrogen atom with a bromine atom. Compare this structure, by attempting to superimpose one structure on top of the other, with the reference model that you constructed in Step 4.

6. Place the reference model aside. On the second model, interchange the bromine with any other hydrogen. Compare this new structure with the reference model by attempting to superimpose the models.

B. Ethane

1. Construct two models of the ethane (C_2H_6) molecule. Sketch the three-dimensional structure of the model.

2. Sketch the shadow of one of the models as in Step 2 of Part A. Held properly, the shadow will have 90° bond angles.

3. Draw the structural formula of ethane. Like the shadow, its 90° bond angles are different than those of the three-dimensional molecule.

4. On each of the two ethane models, replace one of the hydrogen atoms with a bromine atom. Compare their structures by attempting to superimpose the models, and place one of the C_2H_5Br models aside for reference.

5. On the second C_2H_5Br structure, interchange any one of the five hydrogen atoms with the bromine atom. Compare the structure to the reference molecule by attempting to superimpose the models.

6. Repeat Step 5 several times; after each interchange compare the structure to the reference molecule.

7. On one of the C_2H_5Br models, rotate the carbon atoms around the bond joining them so that the atoms assume a "staggered" conformation as shown in Figure 5-2. Draw the shadow of the model as in Step 2 of Part A. Label the bromine atom in the sketch of the shadow.

8. Hold the carbon atom attached to the three hydrogens in a fixed position and rotate the second carbon atom (attached to the bromine) one-third of a turn until a second staggered confirmation is obtained.

9. Holding the structure in the same relative position as in Step 7, sketch the shadow as before. Label the bromine atom.

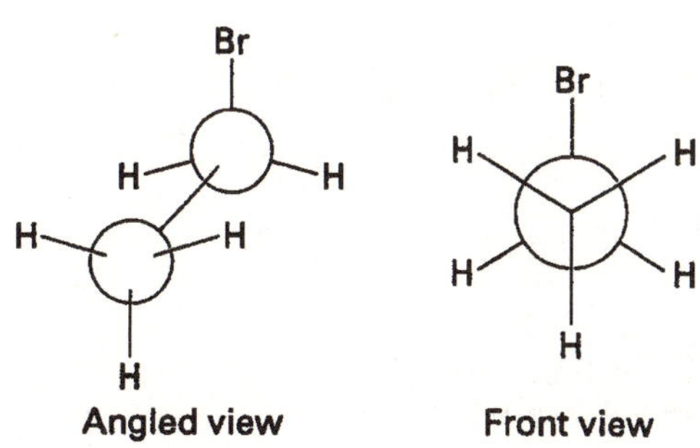

Figure 5-2 Model of bromomethane in "staggered" position.

10. Rotate the carbon atom with the bromine attached to it through another one-third of a turn so that a third staggered conformation is obtained.

11. Sketch the shadow, labeling the bromine atom.

12. Assemble models of as many different isomers of $C_2H_4Br_2$ as you can, comparing their structures by attempting to superimpose the models. Write the structural formula and the IUPAC name of each isomer.

C. Propane

1. Construct a model of a propane (C_3H_8) molecule.

2. Sketch the shadow of the structure as before. Held properly, all bonds in the shadow are at right angles.

3. Draw the structural formula of propane.

4. Construct models of as many different isomers of C_3H_7Br as possible. Compare the structures by attempting to superimpose the models. Write the structural formula of each. Write the IUPAC name of each isomer.

5. Construct models of as many different isomers of $C_3H_6Br_2$ as you can. Write their structural formulas and names.

D. Butane and Pentane

1. Construct models of as many different isomers of C_4H_{10} as you can. Write the structural formula of each and write the IUPAC name of each.

2. Construct models of as many different isomers of C_5H_{12} as possible. Write the structural formula and name of each isomer.

E. Ethylene (Ethene)

1. Construct a model of ethylene (C_2H_4). Use two bent-pegs or two bendable spring-pegs (depending on the type of model kit you are using) between the carbon atoms to represent the double bond. Write the structural formula of ethylene. Try to rotate the two carbon atoms as you did in Step B7.

2. Construct models of as many different isomers of $C_2H_2Br_2$ as possible. Write their structural formulas and names.

F. Cyclopentane

1. Construct a model of cyclopentane (C_5H_{10}). Write its structural formula. Attempt to rotate any two carbon atoms.

2. Construct models of as many different cyclic isomers of C_5H_9Br as possible. Write the structural formula and name of each.

3. Construct models of as many different cyclic isomers of $C_5H_8Br_2$ as possible. Write the structural formula and name of each.

II. PROPERTIES OF HYDROCARBONS

A. Solubility

1. Place 8 drops of the following hydrocarbons into 3 numbered test tubes as follows:
   No. 1: heptane
   No. 2: 1-hexene
   No. 3: toluene

2. Add 3 mL of deionized water to each test tube; cork and shake vigorously. Classify the compounds as insoluble (I) or soluble (S) in water.

3. Disposal: Pour the liquids into the "Waste Hydrocarbon" container.

CAUTION: Heptane, 1-hexene, and toluene are flammable. As a general rule, consider all organic compounds to be combustible. The vapors of many organic compounds are believed to be harmful to your health; physicians may advise pregnant women to avoid breathing vapors of organic compounds. To reduce the concentration of vapors in the room, the hydrocarbons used in this experiment should be handled only in a fume hood.

B. Combustion

WE RECOMMEND THAT STEP B OF THIS EXPERIMENT IS CONDUCTED BY AN INSTRUCTOR FOR DEMONSTRATION PURPOSES ONLY, DUE TO THE FIRE DANGER INVOLVED WITH THESE PROCEDURES

1. **Perform this test in a fume hood. Perform the test in an area far from open bottles of hydrocarbons.**

2. Place about 5 drops of heptane in an evaporating dish. With a burning splint **at arm's length**, ignite the heptane. Observe the color of the flame, smoke, and residue (if any).

3. Repeat Steps 1–2 with 1-hexane and with toluene.

C. Test for Unsaturation

The bromine solution may burn and blister the skin. Bromine vapors are irritating to eyes and lungs. Bromine solutions should be handled only in a fume hood. Wear disposable laboratory gloves to handle this solution.

1. **Perform this test in a fume hood.**

2. Pour approximately 1 mL of heptane into a clean, dry test tube (to a height of approximately 1 cm). Note its color.

3. Note the red-brown color of bromine solution (bromine dissolved in dichloromethane). Add the bromine solution drop by drop into the test tube until a faint color of bromine persists. **CAUTION: Gently shake the test tube after the addition of each drop.** Count the number of drops required to produce a faint bromide color. **CAUTION: Do not add more than 20 drops.**

4. Repeat Steps 1–3 with 1-hexene and with toluene.

5. Disposal: Pour the 3 hydrocarbon solutions from the test tubes into the "Halogenated Solvent Waste" container. Clean the test tubes with a brush and detergent.

Name ______________________________ Section __________ Date ____________

Prelab Questions for Experiment 5

________ 1. Two structures are identical when
a. they have the same molecular formula
b. they have the same combustion products
c. they can be superimposed on each other

________ 2. Toluene is classified as
a. an alkane
b. an alkene
c. an alkyne
d. an aromatic hydrocarbon

________ 3. Cyclopentane is classified as
a. an unsaturated compound
b. a saturated compound
c. an alkene
d. an alkyne

________ 4. Several drops of a bromine solution were added to a liquid. The red-brown color of the bromine immediately disappeared. The compound may have been
a. a saturated hydrocarbon
b. a cycloalkane
c. an aromatic hydrocarbon
d. an alkane
e. an alkene

5. a. Write the structural formula of heptane.

b. What is the molecular formula of heptane?

6. a. Write the structural formula of 1-hexene.

b. What is the molecular formula of 1-hexene?

7. a. Write the structural formula of toluene.

b. What is the molecular formula of toluene?

8. a. In the molecular model set, what type of bond is represented by a long-peg?

b. A short-peg?

____

c. Two bent-pegs?

____

9. Complete and balance the following equations:

a.

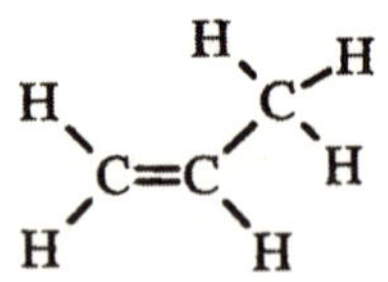 + Br_2 ⟶

b. C_5H_{12} + 8 O_2 →

Name ______________________________ Section ____________ Date ______________

Report for Experiment 5

I. MOLECULAR MODELS: ISOMERISM

A. Methane

1. Sketch of the three-dimensional molecule:

2. a. Sketch of the methane shadow: b. Structural formula of methane:

3. Are all the hydrogens in the methane **equivalent**; that is, do you obtain the same structure by replacing any one of the hydrogens with bromine?

____

B. Ethane

1. a. Sketch of the three-dimensional ethane molecule:

   b. Sketch of the ethane shadow: b. Structural formula of ethane:

2. Are all the hydrogens in ethane equivalent; that is, do you obtain the same molecule by replacing any hydrogen with bromine?

____

3. Sketches of C_2H_5Br shadows:

a. Step 7

b. Step 9 (after one-third of a turn)

c. Step 11 (after another one-third of a turn)

d. Do the above sketches of C_2H_5Br represent the same or different structures?

e. Do the following structural formulas represent the same or different molecules?

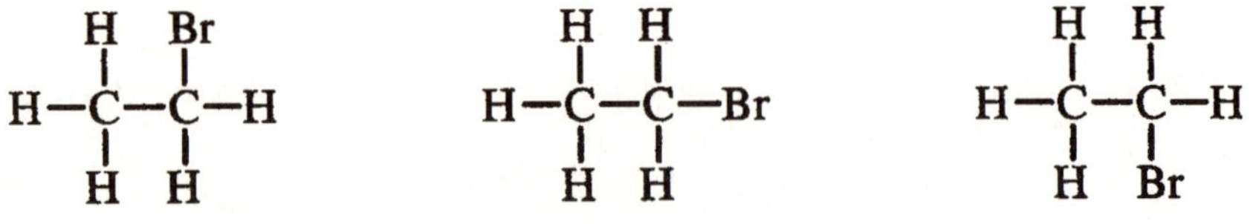

4. Structural formulas and names of isomers of $C_2H_4Br_2$:

Name ______________________ Section ____________ Date ______________

C. Propane

1. a. Sketch of the propane shadow:

   b. Structural formula of propane:

2. Structural formulas and names of isomers of C_3H_7Br:

3. Structural formulas and names of isomers of $C_3H_6Br_2$:

D. Butane and Pentane

1. Structural formulas and names of isomers of C_4H_{10}:

2. Structural formulas and names of isomers of C_5H_{12}:

E. Ethylene (Ethene)

1. a. Structural formula of ethylene:

   b. Do the carbon atoms have freedom of rotation?

2. Structural formulas and names of isomers of $C_2H_2Br_2$:

F. Cyclopentane

1. a. Structural formula of cyclopentane:

   b. Do the carbon atoms have freedom of rotation?

2. Structural formulas and names of cyclic isomers of C_5H_9Br:

3. a. Structural formulas and names of cyclic isomers of $C_5H_8Br_2$:

   b. For geometric (*cis-trans*) isomerism to occur, must a double bond be present in the structure? ______________________________

Name ______________________ Section ____________ Date ____________

II. PROPERTIES OF HYDROCARBONS

A. Solubility

1. Data:

Hydrocarbon	*Solubility in Water (I or S)*
Heptane	
1-Hexene	
Toluene	

2. In what type of solvent, polar or nonpolar, are hydrocarbons soluble?

____

B. Combustion

1. Data:

Hydrocarbon	*Color of Flame*	*Amount of Smoke*	*Amount of Residue*
Heptane			
1-Hexene			
Toluene			

2. Which class of hydrocarbon (alkane, alkene, or aromatic) is the cleanest burning fuel?

____

3. All classes of hydrocarbons burn completely when premixed with sufficient oxygen (as in a Bunsen burner with the air inlet open). Write a balanced equation for the complete combustion of each hydrocarbon. You may use molecular formulas rather than structural formulas.

____

____

____

C. Test for Unsaturation

1. Data:

Hydrocarbon	*Number of Br_2 Drops*
Heptane	
1-Hexene	
Toluene	

2. a. Which of the hydrocarbons reacted with bromine?

____

b. Write a balanced equation for each hydrocarbon that reacted with bromine. Use structural formulas (not molecular formulas) in the equation(s).

Related Questions for Experiment 5

Draw structural formulas for all the isomers of C_6H_{14}. Name each isomer.

Experiment 6
Properties of Gases

Reference: *General, Organic, and Biological Chemistry: An Integrated Approach, Chapter 5.*

Purpose: You will prepare carbon dioxide gas and observe its properties. You will investigate Boyle's Law, Henry's Law, and Charles' Law. You will use the Pressure-Temperature Law to determine the value of absolute zero.

Introduction

CARBON DIOXIDE GAS: PREPARATION AND PROPERTIES

Carbon dioxide gas is one of the endproducts of digesting food and it is in the air we exhale, but our exhaled breath is only 4% CO_2. The atmosphere we breathe contains 0.03% CO_2. The presence of carbon dioxide in the blood stimulates breathing. For this reason, carbon dioxide is added to oxygen in artificial respiration and to the gases used in anesthesia.

Carbon dioxide is also produced when any organic (carbon-containing) fuel burns. Fuels such as gasoline, coal, natural gas, wood and paper all produce CO_2 when they burn. Burning too many of these fuels may add too much carbon dioxide to our atmosphere—that can lead to overheating of the earth in a process known as global warming. Plants remove carbon dioxide from the air and make food from it in the photosynthesis process. Carbon dioxide is also removed from the atmosphere by dissolving in sea water.

Carbonated beverages such as soda contain dissolved carbon dioxide. In baking cookies or cakes, a chemical reaction produces CO_2 from either baking soda or baking powder. The carbon dioxide gas produces the tiny holes in cookies and cakes. In baking bread or pizza dough, yeast produces carbon dioxide from sugar—the CO_2 makes the dough rise.

Baking soda and acids (such as vinegar) react to form carbon dioxide. Baking soda is sodium bicarbonate, $NaHCO_3$. Vinegar contains acetic acid, $HC_2H_3O_2$, in water. The production of this gas causes the fizzing that you will see in the reaction.

$$NaHCO_3\,(s) + HC_2H_3O_2\,(aq) \rightarrow CO_2\,(g) + NaC_2H_3O_2\,(aq) + H_2O\,(l)$$

In the above equation (*s*) represents a solid, (*g*) is a gas, (*l*) is a liquid, and (*aq*) is an aqueous solution. An **aqueous** solution is a mixture in which something is dissolved in water.

Carbon dioxide gas can extinguish a fire. It does this by reducing the amount of oxygen in the air around the burning object. Oxygen in the air is necessary for a fire and when this oxygen is removed, the fire goes out.

Although carbon dioxide is a gas at room temperature, it is a solid at temperatures below –78.5°C (–109.3°F). Solid carbon dioxide is called "dry ice" since it changes from solid to gas without ever forming a liquid, a process called **sublimation**.

$$CO_2\,(s) \xrightarrow{\text{heat}} CO_2\,(g)$$

Manufacturing carbon dioxide is big business. Billions of tons of carbon dioxide are produced each year in the United States making it a comercially important material.

CAUTION! Do not touch dry ice with your bare skin; the extreme cold can severely damage the skin.

THE GAS LAWS

Not all gases are colorless, odorless, solidify at –78.5°C and extinguish fires; these are properties of carbon dioxide gas. All gases do share certain properties described by the gas laws. All gases obey Boyle's Law, Henry's Law and Charles' Law if the gases are at normal or high temperature and normal or low pressure.

BOYLE'S LAW: THE PRESSURE-VOLUME LAW

Boyle's Law states that the volume of a gas in inversely proportional to the pressure when the temperature remains constant. In simple terms, the volume of a gas increases as the pressure decreases; conversely, the volume of a gas decreases with an increase in pressure. This law is qualitatively illustrated in the first portion of this experiment.

A small, sealed balloon is placed into a suction flask. See Figure 6-2. A water aspirator is used to decrease the pressure of a gas (air) in the suction flask and in the balloon. The water aspirator is attached to the cold water faucet. When water flows rapidly through the aspirator, suction is created in the side arm. The aspirator works on Bernoulli's Principle, which states that as the velocity of a liquid or gas is increased, its pressure is decreased. An aspirator cannot provide a pressure lower than the vapor pressure of the flowing water. The lowest pressure of cold water is about 10–20 mm Hg. A trap is used to prevent water from entering the flask containing the balloon. Thick-walled rubber vacuum tubing is needed to prevent it from collapsing due to the pressure difference. When the water is turned on, air is evacuated from the suction flask and the pressure exerted on the balloon decreases. To remove the suction, first disconnect the vacuum tubing, and then turn off the water. Turning off the water first may cause water to enter the trap.

In the second portion of the experiment, the physiological application of Boyle's Law is illustrated with a lung demonstration model. The model consists of two rubber balloons attached to a Y-tube inserted in a rubber stopper in the top of a bell jar. The rubber stopper end of the Y-tube is open to the air—this allows air to enter and leave the balloons. A thin sheet of rubber is stretched across the bottom of the bell jar. The stopper and rubber sheet seal the bell jar air-tight. No air can enter or leave the bell jar; air already present in the bell jar can change in volume and pressure. See Figure 6-3.

HENRY'S LAW: THE PRESSURE-SOLUBILITY LAW

According to **Henry's Law**, the solubility of a gas in a liquid is directly proportional to the pressure of the gas on the liquid. Increasing the pressure increases gas solubility; decreasing the pressure decreases its solubility. An unopened bottle of a carbonated beverage contains a large amount of carbon dioxide (CO_2) gas dissolved in the aqueous solution. While dissolved, the gas is invisible—the mixture is homogenous. An open bottle of carbonated beverage has bubbles of carbon dioxide gas rising to the surface. These bubbles of gas are not dissolved. Opening the bottle decreases the pressure and decreases the water solubility of carbon dioxide.

CHARLES' LAW: THE VOLUME-TEMPERATURE LAW

Charles' Law states that the volume of a gas varies directly with the absolute temperature, provided that the pressure remains constant. That is, the volume of a gas will increase when the absolute temperature increases, if the pressure remains unchanged.

THE PRESSURE-TEMPERATURE LAW AND ABSOLUTE ZERO

Temperature is a measure of the average kinetic energy of the molecules in a sample of matter. At room temperature, molecules are moving very rapidly. When molecules slow down, their temperature decreases. Theoretically, at some very cold temperature the motion of the molecules would slow to zero and the molecules would be totally motionless. This temperature, called **absolute zero**, is equal to –273.15°C or 0 K. Scientists have recently achieved a temperature of 0.000000002 K, very close to absolute zero. Modern theory says that absolute zero can never be reached; molecular motion can never completely cease.

In a rigid container (one in which the volume remains constant), increasing the temperature will increase the pressure of the gas. Lowering the temperature will lower the pressure. This relationship is sometimes known as the **Pressure-Temperature Law**, which states that the pressure of a gas varies with the absolute temperature provided that the volume remains constant.

In this demonstration, a pressure gauge has been attached to a hollow steel ball. See Figure 6-1. The hollow ball contains a gas. Any gas (such as nitrogen, oxygen, or helium) will work. Our ball is filled with air which is a mixture of mainly nitrogen, oxygen, and argon gases. The ball will be immersed in liquids at three different temperatures and the pressure in the ball will be measured. The data points will be plotted on graph paper and a best-fitting straight line will be drawn. The line will be extended to lower temperatures and lower pressures. One point on the line corresponds to zero pressure. Absolute zero is the temperature at which the pressure would be zero.

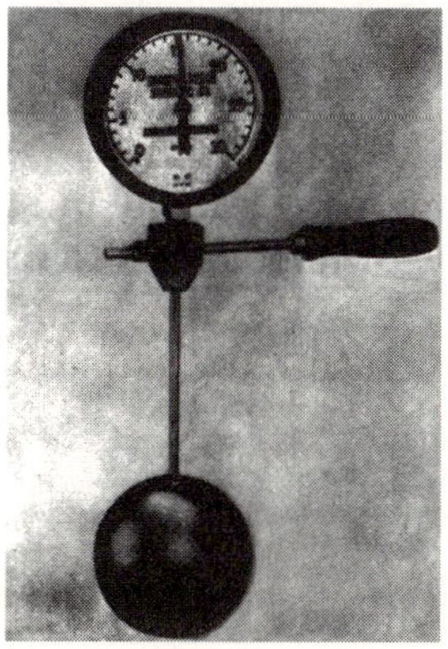

Figure 6-1 Apparatus to demonstrate absolute zero.

Procedure

I. CARBON DIOXIDE GAS: PREPARATION AND PROPERTIES

1. Make a candle holder by cutting a piece of corrugated cardboard to a size of about 5 cm × 5 cm. Use a sharp awl to pierce a small hole in the center of the cardboard.

2. Insert a birthday candle through the hole. The candle should stand upright on the cardboard base.

3. Place the cardboard and candle in the bottom of a 400-mL beaker. Then cover the bottom of the beaker with about 10 g of baking soda (sodium bicarbonate, $NaHCO_3$). It is not necessary to accurately measure the quantity of sodium bicarbonate.

4. Obtain about 20 mL of vinegar (5% acetic acid, $HC_2H_3O_2$) in a 50-mL beaker.

5. Holding a lighted match with tongs, light the candle.

6. Slowly pour the vinegar down the inside edge of the beaker, NOT ON THE CANDLE! Observe what happens. If the candle goes out, try to re-light it.

II. DRY ICE

1. Using tongs or insulated gloves, place a small piece (about the size of a marshmallow) of dry ice in a quart-size zip-close plastic bag. **CAUTION: Do not touch the dry ice with your bare skin; the extreme cold can cause serious injury.**

2. Flatten the bag to remove as much air as possible and seal the bag shut.

3. Allow the dry ice to slowly warm up. Observe.

III. BOYLE'S LAW: THE PRESSURE-VOLUME LAW

A. Volume-Pressure Relationship

1. Slightly inflate a small balloon to a size so that it can be placed inside a 250-mL suction flask. Knot the end tightly to prevent air from escaping.

2. Set up the apparatus as illustrated in Figure 6-2 but do not insert the stopper. Securely attach one length of rubber vacuum tubing from the stem of the flask to the trap. The second length of vacuum tubing connects the trap to the side arm of the water aspirator. The vacuum tubing has thick walls to prevent its collapse.

3. Run the water at full force and place the palm of your hand on top of the suction flask. If the aspirator is working properly you should feel a strong suction due to the low pressure inside the suction flask.

4. Without turning off the water, remove your hand and tightly insert the stopper into the suction flask. Observe the behavior of the balloon in the flask.

5. Detach the vacuum tubing from the aspirator and allow air to enter the flask. Observe the behavior of the balloon.

6. Shut off the water. Save the suction flask set-up for Part IV.

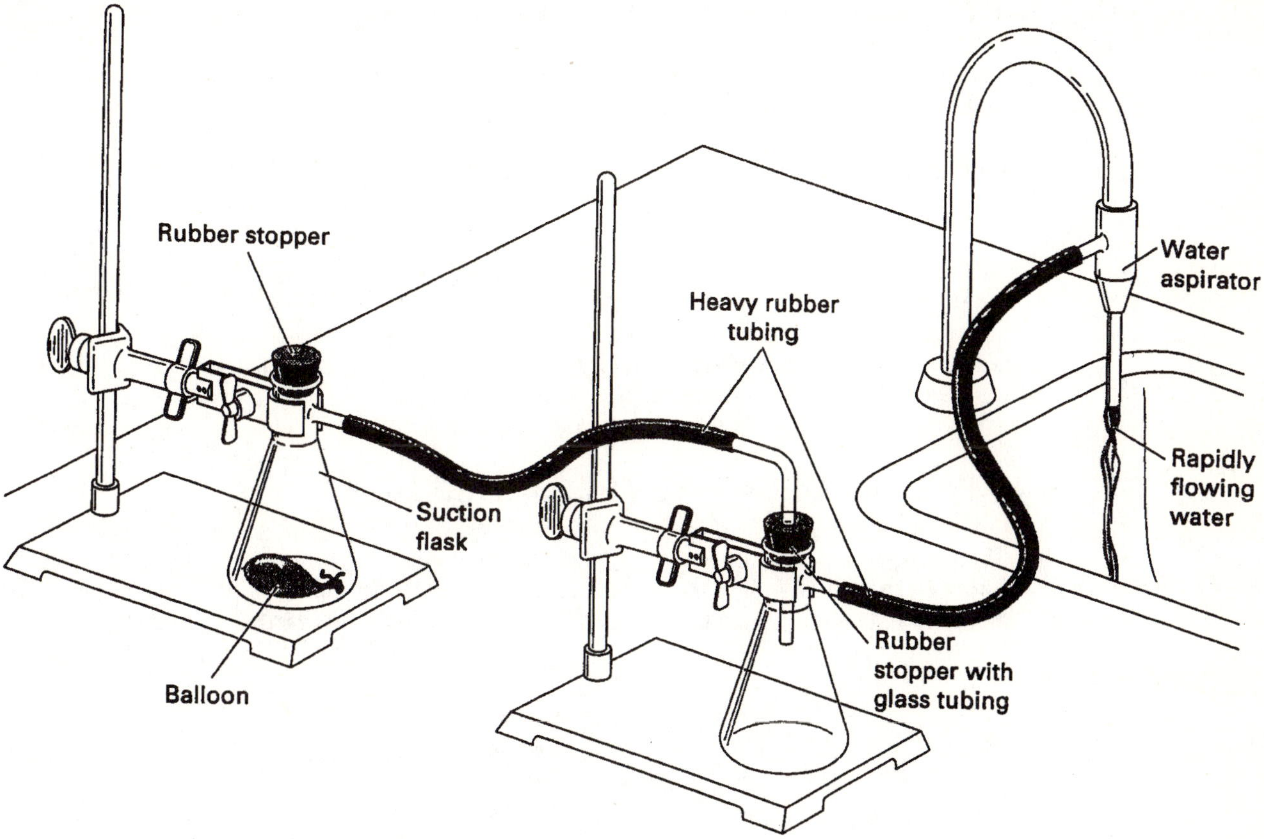

Figure 6-2 The setup for studying volume-pressure relationship.

B. Physiological Application of Boyle's Law

1. Obtain a "lung demonstration model" from your instructor.

2. Pull out the thin rubber diaphragm as illustrated in Figure 6-3. If you have long fingernails, take care not to puncture the rubber. Observe the behavior of the balloons.

3. Release the rubber diaphragm and allow it to return to its original position. Observe the behavior of the balloons.

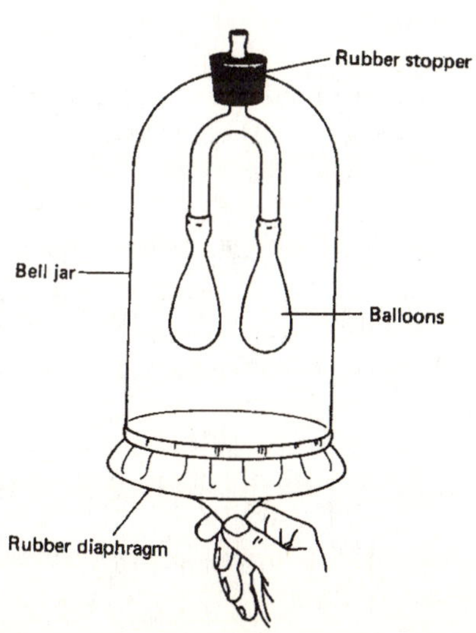

Figure 6-3 Lung demonstration model.

IV. HENRY'S LAW: THE PRESSURE-SOLUBILITY LAW

1. Remove the balloon used in Procedure III.A. from the suction flask. Empty any water that may have entered the trap.

2. Pour 25 mL of carbonated water into the suction flask.

3. Connect the vacuum tubing to the water aspirator.

4. Run the water at full force and observe the behavior of the carbonated water.

5. Disconnect the vacuum tubing and observe the behavior of the carbonated water. Shut off the water. Rinse out and return the suction flask.

V. CHARLES' LAW: THE VOLUME-TEMPERATURE LAW

1. Obtain an 8-cm piece of 1.5-mm capillary tubing that is sealed at one end. A small drop of colored water has been inserted into the capillary so as to trap some air in the closed end. See Figure 6-4. The capillary tube must be handled very carefully in all succeeding steps.

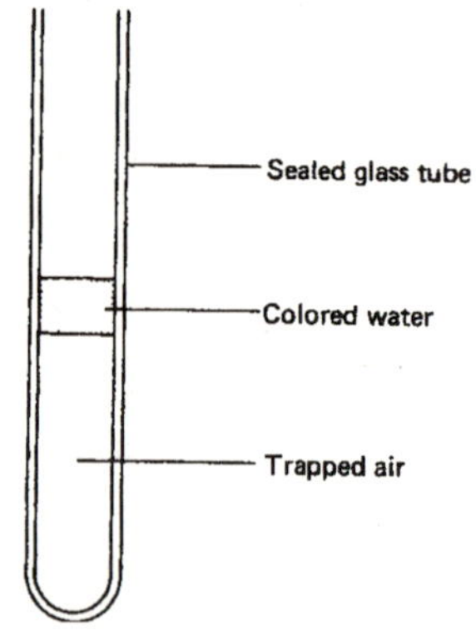

Figure 6-4 Apparatus for demonstrating Charles' Law.

2. Carefully draw a line with a wax marking pencil or marking pen at the lower edge of the colored water.

3. Prepare an ice-water bath in a 150-mL beaker.

4. Immerse the tubing in the ice-water bath, open end up. Observe the effect of temperature on the volume of air trapped below the colored water. Allow at least 1 minute for the temperature of the gas to cool to the temperature of the ice-water bath.

5. Using your thumb, mark the position of the lower edge of the colored water and remove the tube from the ice bath.

6. Dry the tubing and use a wax marking pencil or marking pen to mark your thumb position on the tube.

7. Repeat Steps 4, 5, and 6 with a warm water bath, heated to approximately 60°C.

8. Remove the wax pencil lines by wiping the tube with a paper towel or with a small amount of acetone on a paper towel. Return the capillary tubing with the colored water still inside.

VI. THE PRESSURE-TEMPERATURE LAW AND ABSOLUTE ZERO (Instructor's Demonstration)

1. Your instructor will immerse the bulb of the apparatus to demonstrate absolute zero in three solutions at three different temperatures. The recommended solutions are ice water (0°C), boiling water (100°C), and dry ice-isopropyl alcohol (–78.5°C). Record the temperature and pressure for each of the solutions.

2. Plot the three data points on the graph in the Report Section (see Figure 6-5).

3. Use a ruler and draw one single straight line which comes closest to passing through all three data points.

4. Extend the straight line to lower pressures and lower temperatures. Determine the temperature at which the pressure is zero—this temperature is absolute zero.

Name ______________________ Section ____________ Date ____________

Prelab Questions for Experiment 6

1. What safety precautions are necessary when handling dry ice?

____

2. Describe the order of steps in shutting off a water aspirator.

____

____

3. How does the volume of a gas change when the pressure is increased?

____

4. How does the volume of a gas change when the temperature is increased?

____

5. How does the water-solubility of a gas change when the pressure is increased?

____

6. In Part III. B., which parts of the human body are represented by the following pieces of the lung demonstration model:

   a. The two rubber balloons ______________________

   b. The Y-tube ______________________

   c. The open end of the Y-tube ______________________

   d. The bell jar ______________________

   e. The thin sheet of rubber ______________________

Name ______________________ Section ____________ Date ____________

Report for Experiment 6

I. CARBON DIOXIDE GAS: PREPARATION AND PROPERTIES

1. What is the evidence that baking soda and vinegar react to produce a gas?

____

2. What properties does carbon dioxide gas have (color, effect on fires, etc.)?

____

____

3. Were you able to relight the candle? ____________

4. Is there any evidence that carbon dioxide gas is lighter or heavier than air? If so, what?

____

____

II. DRY ICE

1. What happened to the dry ice in the bag?

____

2. How could you test the gas in the bag to show it was carbon dioxide?

____

____

III. BOYLE'S LAW: THE PRESSURE-VOLUME LAW

A. Volume-Pressure Relationship

1. a. What happened to the balloon when you decreased the pressure in the suction flask?

____

b. Explain your observations in *a*.

____

____

2. a. What happened to the balloon when you detached the vacuum tubing from the aspirator?

____

b. Explain your observations in *a*.

____

____

B. Physiological Application of Boyle's Law

1. a. What happened to the balloons when you pulled out the rubber diaphragm?

____

b. Explain your observations in *a*.

____

____

2. a. What happened to the balloons when the diaphragm returned to its original position?

____

b. Explain your observations in *a*.

____

____

IV. HENRY'S LAW: THE PRESSURE-SOLUBILITY LAW

1. a. What happened to the carbonated water when the pressure in the suction flask decreased?

____

b. Explain your observations in *a*.

____

____

2. a. What happened to the carbonated water when the vacuum tubing was disconnected from the aspirator?

____

b. Explain your observations in *a*.

____

____

V. CHARLES' LAW: THE VOLUME-TEMPERATURE LAW

What happened to the volume of trapped air as the temperature increased?

____

____

VI. THE PRESSURE-TEMPERATURE LAW AND ABSOLUTE ZERO (Instructor's Demonstration)

1. a. Data: (Remember to include units with each quantity):

Temperature	Pressure
____________	____________
____________	____________
____________	____________

2. a. On your graph, what is the indicated value of absolute zero?

____

b. How does your experimental value compare with the accepted value of absolute zero?

____

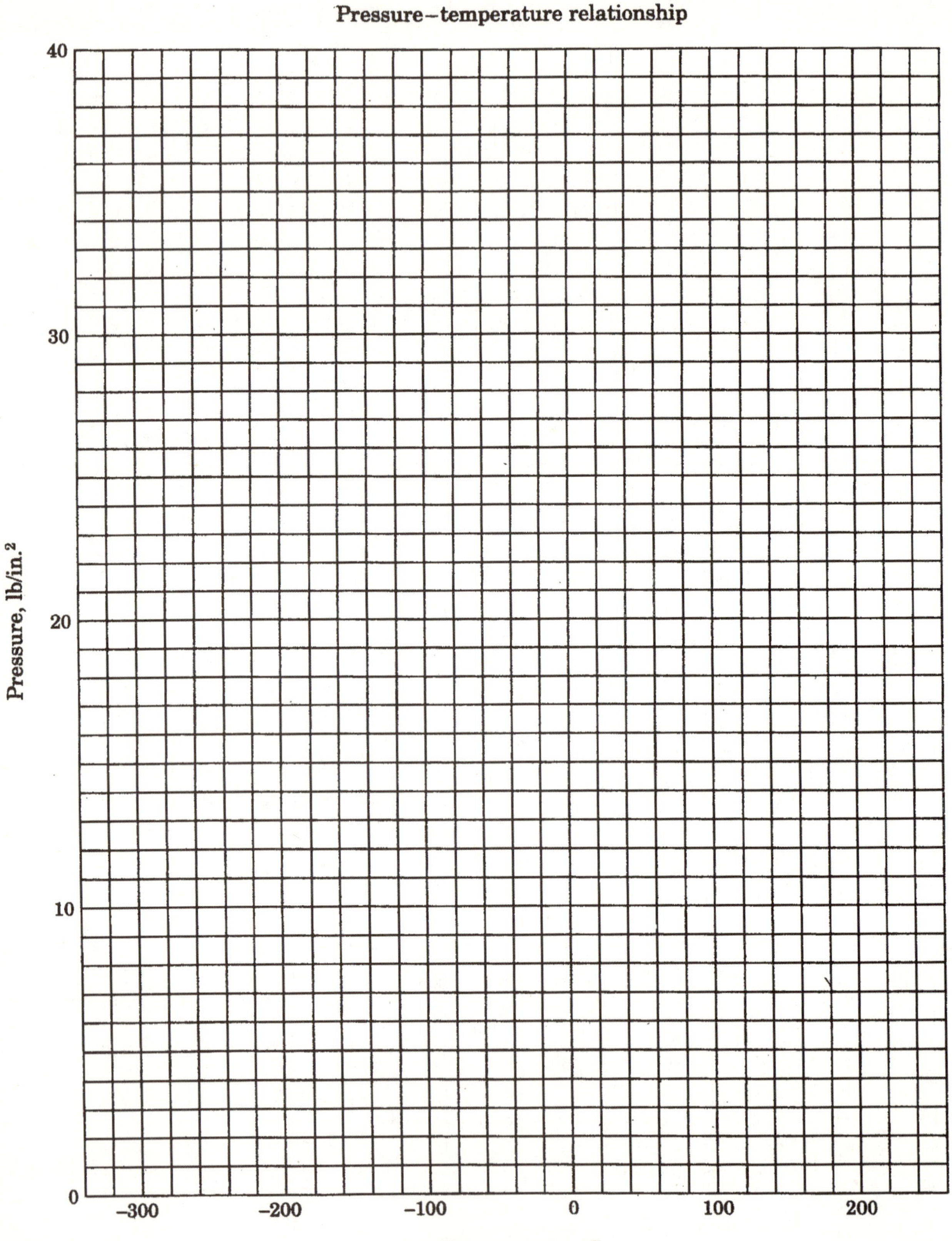

Figure 6-5 The relationship between pressure and temperature of a gas. Absolute zero is the temperature at which the pressure is zero.

Name ______________________ Section ____________ Date ______________

Related Questions for Experiment 6

1. Use the *CRC Handbook of Chemistry and Physics* to find the solubility of carbon dioxide in water. It will be in the section "Physical Constants of Inorganic Compounds" of the 75th (1994–95) edition or earlier editions. This information can also be found in the Merck Index.

   a. Solubility in cold water (0°C) ______________ cm^3 CO_2 per 100 cm^3 H_2O

   b. Solubility in hot water (20°C) ______________ cm^3 CO_2 per 100 cm^3 H_2O

   c. Solubility in cold water (0°C) ______________ g CO_2 per 100 cm^3 H_2O

   d. Solubility in hot water (60°C) ______________ g CO_2 per 100 cm^3 H_2O

   e. This difference seen in solubility at different temperatures is typical of gases. How does increasing temperature affect the solubility of a gas in water?

   ____

   ____

2. "Thermal pollution" is the term used to describe hot water added to a lake or river by an electrical generating plant or other sources. Why is increased water temperature an environmental concern?

   ____

   ____

3. a. What is "the bends"?

   ____

   ____

   b. Which portion of this experiment illustrates the cause of the bends?

   ____

4. Explain why supplying a mixture of helium and oxygen gases rather than compressed air to sea divers prevents "the bends."

   ____

   ____

5. An old method of treating "the bends" was to pack the diver in ice. Explain why lowering the body temperature served as an emergency treatment.

____

____

6. In a tragic 1989 accident, a cargo door of an airplane fell off as the plane flew at 22,000 feet over the South Pacific. As the door fell away, it ripped out a large hole in the passenger cabin. Nine passengers were swept out of the opening and many more were hurled toward the opening but managed to stay inside the plane. Use the gas laws to explain why the passengers were affected in this way.

____

____

7. Explain why aerosol cans should be stored in a cool place.

____

____

8. Explain why rapid ascent in an elevator may cause the volume of intestinal gas to increase and cause intestinal cramps.

____

____

9. In 1703, the French physicist Guillaume Amontons invented the gas thermometer. Explain how you could use a marking pen to convert the absolute zero demonstrator into a gas thermometer.

____

____

10. Liquid mercury thermometers are usable only between –40°C and 360°C. Why can't they be used at lower or higher temperatures?

____

____

Experiment 7

Stoichiometry: Mole Relationships

Reference: *General, Organic, and Biological Chemistry: An Integrated Approach, Chapters 3 and 6.*

Purpose: You will establish the validity of a balanced equation by experimentally determining the molar ratio of reactants and products in a chemical reaction.

Introduction

REACTIONS AND EQUATIONS

Propane (C_3H_8) is a gaseous fuel sold in steel cylinders and used in camping stoves, propane torches, and gas barbecue grills. The combustion of propane is an example of a chemical reaction. A chemical reaction and a chemical equation are not the same. You cannot see this reaction until you have lit your camp stove. A chemical equation is a symbolic representation of the chemical reaction. The equation for the reaction is:

$$C_3H_8 \ + \ 5\ O_2 \longrightarrow 3\ CO_2 + 4\ H_2O$$

Chemical reactions are automatically balanced by nature—mass is never created nor destroyed in a chemical reaction. Chemical equations are not automatically balanced. A chemist must balance chemical equations by inserting coefficients in front of the reactants and products.

The coefficients of a balanced equation can be interpreted in two ways:

1. **Coefficients as number of molecules.** The equation above can be read as:

   "one molecule of propane reacts with five molecules of oxygen to produce three molecules of carbon dioxide and four molecules of water."

2. **Coefficients as number of moles.** Since molecules are so tiny and since so many molecules are involved in a reaction, chemists count molecules in a larger unit called a mole. One mole of any substance is 6.02×10^{23} molecules. Therefore the equation can also be read as:

   "one mole of propane reacts with five moles of oxygen to produce three moles of carbon dioxide and four moles of water."

The actual number of moles of propane that burn in the camp stove depends on the length of time the stove is lit. No matter how long the stove is lit, the ratio of moles of oxygen consumed to moles of propane consumed will remain at 5:1. **The coefficients of the equation represent the simplest whole number ratios of moles of reactants and products.**

BALANCING EQUATIONS BY TRIAL-AND-ERROR

Aluminum metal reacts with hydrochloric acid to produce aluminum chloride and hydrogen gas. The unbalanced equation can be written as:

$$Al + HCl \longrightarrow AlCl_3 + H_2$$

The equation can be balanced by trial-and-error. Usually it is best to balance hydrogen and oxygen last, after all other elements have been balanced. Let's begin by noticing that aluminum is already balanced. Balance chlorine by adding a coefficient of 3 HCl.

$$Al + 3\,HCl \longrightarrow AlCl_3 + H_2$$

There are three hydrogen atoms on the left and two on the right side. We need six hydrogen atoms on each side to balance the equation.

$$Al + 6\,HCl \longrightarrow AlCl_3 + 3\,H_2$$

The chlorine atoms have become unbalanced. The left side has 6 chlorine atoms and the right side has 3 chlorine atoms. To balance chlorine place a coefficient of 2 in front of $AlCl_3$.

$$Al + 6\,HCl \longrightarrow 2\,AlCl_3 + 3\,H_2$$

The final step is balanced since there are two aluminum, six hydrogen, and six chlorine atoms on each side of the equation. Mass is conserved; no atoms are created nor destroyed in the chemical reaction.

BALANCING EQUATIONS FROM EXPERIMENTAL DATA

We could experimentally verify the coefficients of the equation by reacting two moles of aluminum with six moles of hydrochloric acid. If the products were collected and measured there would be two moles of aluminum chloride and three moles of hydrogen produced in the reaction.

Actually we can begin with any amount of aluminum and still demonstrate the validity of the equation. Consider the following data collected by a student. She measured the masses of aluminum and hydrochloric acid that were used in her experiment and the masses of aluminum chloride and hydrogen that were produced.

Substance	*Mass*
Al	103 g
HCl	416 g
$AlCl_3$	507 g
H_2	11.5 g

Let's see how her experimental data verifies the validity of the balanced equation.

Step 1: Calculate the number of moles of each substance. The formula weight of a substance can be found by adding the atomic weights of every atom in the formula.

Formula weight of Al = 26.98 mu

Molecular weight of HCl = 1.01 + 35.45 = 36.46 amu

Formula weight of $AlCl_3$ = 26.98 + (3 × 35.45) = 133.33 amu

Molecular weight of H_2 = 2 × 1.01 = 2.02 amu

The mass of 1 mole of a substance is equal to the formula weight or molecular weight expressed in units of grams. The mass of one mole (the molar mass) can be used as a conversion factor.

$$\frac{103\text{ g A1}}{1} \times \frac{1\text{ mole A1}}{26.98\text{ g A11}} = 3.82\text{ mole A1}$$

$$\frac{413\text{ g HC1}}{1} \times \frac{1\text{ mole HC1}}{39.46\text{ g HC1}} = 11.4\text{ mole HC1}$$

$$\frac{507\text{ g A1C1}_3}{1} \times \frac{1\text{ mole A1C1}_3}{133.33\text{ g A1C1}_3} = 3.80\text{ mole A1Cl}_3$$

$$\frac{11.5\text{ g H}_2}{1} \times \frac{1\text{ mole H}_2}{2.02\text{ g H}_2} = 5.69\text{ mole H}_2$$

The coefficients of an equation represent the number of moles of each substance. Therefore we might try to write the equation for the reaction as:

$$3.82\text{ Al} + 11.4\text{ HCl} \longrightarrow 3.80\text{ AlCl}_3 + 5.69\text{ H}_2$$

This is incorrectly written since **the coefficients must be the smallest whole number molar ratios of the substances.**

Step 2: Find the molar ratios. To find the ratios of moles of the reactants and products, divide each number by the smallest number. The smallest number is 3.80.

Substance	*Molar ratios*
Al	3.82 ÷ 3.80 = 1.01
HCl	11.4 ÷ 3.80 = 3.00
$AlCl_3$	3.80 ÷ 3.80 = 1.00
H_2	5.69 ÷ 3.80 = 1.50

We could use the molar ratios to write the equation as:

$$1.01\text{ Al} + 3.00\text{ HCl} \longrightarrow 1.00\text{ AlCl}_3 + 1.50\text{ H}_2$$

This is still incorrect since the coefficients of the equation must be the smallest whole number ratios. The coefficient 1.01 Al is very close to a whole number but the coefficient 1.50 H_2 is not close to a whole number.

Step 3: Find the whole number molar ratios. If the molar ratios are not whole numbers, multiply each coefficient by 2 to obtain whole numbers. If 2 does not produce whole numbers, continue using the next higher integer (3, 4, 5) until roundable to a whole number.

Substance	*Mass*	*Number of moles*	*Molar Ratios*	*Whole Number Molar Ratios*
Al	103 g	3.82 mol	1.01	2.02
HCl	416 g	11.4 mol	3.00	6.00
$AlCl_3$	507 g	3.80 mol	1.00	2.00
H_2	11.5 g	5.69 mol	1.50	3.00

These whole number molar ratios are based on experimental data. Since measurements are never exact, the calculated whole numbers are not exactly whole numbers. We have to round off 2.02 to 2 and use it as the coefficient of aluminum.

The whole number molar ratios are the coefficients of the balanced equation!

$$2\,Al + 6\,HCl \longrightarrow 2\,AlCl_3 + 3\,H_2$$

Two moles of aluminum react with six moles of hydrochloric acid to produce two moles of aluminum chloride and three moles of hydrogen. We have experimentally determined the coefficients of the equation by measuring the masses of each reactant and product.

Look again at the masses of the reactants and products.

Mass of reactants = 103 g + 416 g = 519 g
Mass of products = 507 g + 11.5 g = 518.5 g

The Law of Conservation of Mass states that in a chemical reaction, the total mass of the reactants equals the total mass of the products. Since experimental measurements are never exact, small differences in the masses occur but are not significant.

Summary: To find the smallest whole number coefficients of a balanced equation using experimental data:

1. **Find the number of moles of each substance.**

2. **Find the molar ratios by dividing each number of moles by the smallest number of moles.**

3. **If the molar ratios are not whole numbers, multiply the molar ratios by an integer to obtain whole numbers.**

DECOMPOSITION OF POTASSIUM CHLORATE

The reaction you will study is the decomposition of potassium chlorate ($KClO_3$). Potassium chlorate is a white solid that is used as a source of oxygen in explosives and fireworks. When potassium chlorate is heated to a temperature greater than 356°C, it decomposes into potassium chloride and elemental oxygen according to the following unbalanced equation:

$$KClO_3 \longrightarrow KCl + O_2$$

You will heat a weighed sample of potassium chlorate in a crucible. The crucible and sample will be heated in order to expel the oxygen. The residue left in the crucible will be potassium chloride—you will measure the oxygen's mass. There is no direct way to measure the mass of the oxygen gas since it will escape into the air. If the crucible and its contents lose mass during heating, the mass lost must be equal to the mass of the oxygen.

All weighing must be done at room temperature; a hot crucible weighs less than a room temperature crucible just as a hot air balloon weighs less than a room temperature balloon.

CAUTION: Potassium chloride is a powerful oxidizing agent and can cause explosions or fires if heated with combustible materials. Use only clean crucibles and reagent grade potassium chlorate. If spilled, wash the powder with water and allow it to flow down the drain.

Procedure

Make sure you wear safety goggles. Also, make sure all loose clothing and long hair is secured/tied back before lighting the burner.

I. DETERMINING THE NUMBER OF MOLES OF REACTANT AND PRODUCTS

1. Place a clean, dry crucible and cover in a clay triangle. Heat over an intense Bunsen burner flame for about 3 minutes. The tip of the flame's inner cone should just touch the bottom of the crucible.

2. Allow the crucible to cool to room temperature.

3. Determine the mass of the cooled crucible and lid. Measure the mass as accurately as your balance allows, typically ±0.01 g or ±0.001 g. Use the same balance for all subsequent weighings of the crucible. Handle the crucible and cover with crucible tongs for the rest of the experiment; do not use your fingers since they will add moisture and oils to the crucible.

4. Add 1.9–2.1 grams of potassium chloride to the crucible. Determine the mass of the potassium chlorate, crucible, and cover as accurately as your balance allows.

5. Place the partially covered crucible on a clay triangle and gently heat for about 10 minutes. The inner cone of the Bunsen flame should be about 3 inches below the bottom of the crucible. See Figure 7-1. The potassium chlorate should melt and bubble slightly. If the potassium chlorate is not melted, move the inner cone closer to the crucible. Periodically use crucible tongs to remove the cover and monitor the progress of the reaction. **CAUTION: Do not position your face directly above the crucible while examining the contents. Make sure you wear goggles.**

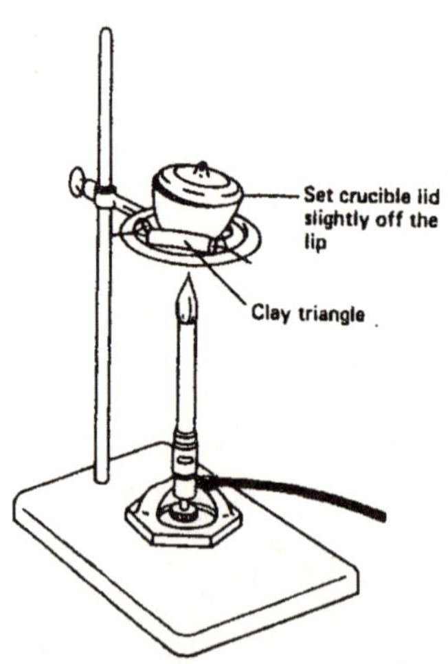

Figure 7-1 Apparatus for heating contents of crucible.

6. After most of the bubbling has stopped, raise the flame of the burner so that the tip of the inner cone just touches the bottom of the crucible. The bottom of the crucible should become red hot. Continue heating for another 10–15 minutes. As potassium chlorate is 770°C, some of the residue may be liquid and some may be solid.

7. After 10–15 minutes of intense heating, allow the crucible to cool.

8. After cooling to room temperature, measure the mass of the crucible, cover, and residue.

9. Place the crucible on the clay triangle once more and heat intensely for another 5 minutes.

10. Cool to room temperature and reweigh the crucible, cover, and residue. Compare the mass of the second heating of the crucible, cover, and residue to the mass of the first heating. If the mass difference is no greater than 0.04 g, no further heating is necessary. If the mass difference is more than 0.04 g, repeat Step 9 until two successive weighings are within 0.04 g. The only way to know if all the potassium chlorate has decomposed is to continue heating and weighing until the mass no longer changes. This is called "heating to a constant mass."

11. If advised by your instructor, repeat Steps 1 through 10 with a second sample.

II. EXAMINATION OF RESIDUE

1. Place three small (10 × 17 mm) numbered test tubes in a rack. Number the tubes with a pencil (not pen) on the white surface or use a stick-on label.

2. Make solutions consisting of the following:

   Test Tube No. 1: A pinch of potassium chlorate ($KClO_3$) in 2 mL of deionized water.

   Test Tube No. 2: A pinch of potassium chloride (KCl) in 2 mL of deionized water.

   Test Tube No. 3: A pinch of residue from the crucible in 2 mL of deionized water.

   You may have to use a spatula to scrape a small quantity of the residue into the test tube. If any of the solids do not dissolve, gently heat it over a Bunsen flame.

3. To each of the three solutions add about 5 drops of 0.1 M silver nitrate ($AgNO_3$). The formation of a white precipitate indicates the presence of a chloride.

$$KCl + AgNO_3 \longrightarrow KNO_3 + AgCl\downarrow$$

The down arrow ($\downarrow$) indicates an insoluble precipitate. The silver chloride precipitate consists of very small pieces of suspended solid. Left undisturbed, the pieces will eventually settle to the bottom of the test tube. All chlorides (e.g., NaCl and $CaCl_2$) give this same precipitate. Since tap water usually contains chloride ions it is important that deionized water be used rather than tap water.

Silver ions, like mercury and lead ions, are heavy metal ions and are toxic. Silver nitrate left on your skin will turn the skin black. Small amounts on the skin are not considered especially dangerous but should be avoided. Swallowing silver nitrate can be fatal. Wash your hands after handling silver nitrate. Dilute solutions of silver nitrate are used to kill bacteria in the eyes of newborn infants. Using the wrong concentration of silver nitrate has caused permanent blindness.

III. CLEANUP

Discard the three solutions and the residue from the crucible in the "Waste Salts" container or as directed by your instructor. Wash your hands.

Name ______________________ Section ____________ Date ____________

Prelab Questions for Experiment 7

1. Why must all masses be determined at room temperature?

____

____

2. After heating and cooling, will the mass of the residue be more than, less than, or the same as the mass of the potassium chlorate originally placed in the crucible? Why?

____

____

3. How can you determine if the decomposition of potassium chlorate is complete?

____

____

4. The following equation represents the reaction of this experiment. Balance it by the trial-and-error method.

$$KClO_3 \longrightarrow KCl + O_2$$

5. When mercury oxide is heated, it decomposes into mercury and oxygen. Balance this equation by trial-and-error.

$$HgO \longrightarrow Hg + O_2$$

6. A student was using the method described in the Procedure for experimentally determining the molar ratios of substances involved in the decomposition of mercury oxide.

She obtained the following data. Complete the chart.

	Mass	*Number of Moles*	*Molar Ratios*	*Whole Number Molar Ratios*
HgO	6.50 g	__________	__________	__________
Hg	6.02 g	__________	__________	__________
O_2	0.48 g	__________	__________	__________

Calculations:

Name ______________________ Section ______________ Date ______________

Report for Experiment 7

I. DETERMINING THE NUMBER OF MOLES OF REACTANT AND PRODUCTS

1. Data: (Remember to include units with the data):

	Trial 1	*Trial 2*
Mass of crucible and cover	________	________
Mass of crucible, cover and $KClO_3$	________	________
Mass of $KClO_3$	________	________
Mass of crucible, cover, and residue (first heating)	________	________
Mass of crucible, cover, and residue (second heating)	________	________
Mass of crucible, cover, and residue (third heating, if necessary)	________	________
Mass of residue (KCl)	________	________
Mass of oxygen	________	________

2. Interpretation of Data:

	Mass	*Number of Moles*	*Molar Ratios*	*Whole Number Molar Ratios*
$KClO_3$	________	________	________	________
KCl	________	________	________	________
O_2	________	________	________	________

Calculations:

3. Are your experimentally determined whole number molar ratios the same as the coefficients of the balanced equation? ______________________________

4. What specific experimental errors might cause a student's experimental whole number ratios to differ from the coefficients of the balanced equation?

____

____

II. EXAMINATION OF RESIDUE

1. What evidence did you find to indicate that the residue was potassium chloride?

____

____

2. What other substances give the same results as potassium chloride when tested with silver nitrate? ______________________________

3. How could you have tested for the presence of oxygen gas as you were heating the potassium chlorate?

____

____

Name ______________________________ Section ____________ Date ______________

Related Questions for Experiment 7

Butane, C_4H_{10}, is a common fuel found in cigarette lighters. When butane burns, it combines with oxygen to produce carbon dioxide and water. A student burned a 5.00 gram sample of butane. Show all calculations in solving the following problems.

1. Write a balanced equation for the combustion of butane.

2. Find the number of moles of butane in 5.00 grams.

3. Find the number of moles of oxygen that will react with 5.00 g of butane.

4. Find the number of grams of oxygen that will react with the butane.

5. Find the number of moles of carbon dioxide produced.

6. Find the number of grams of carbon dioxide produced.

7. Find the number of moles of water produced.

8. Find the number of grams of water produced.

9. Find the total mass of the reactants and compare it to the total mass of the products.

Experiment 8
Properties of Water

Reference: *General, Organic, and Biological Chemistry: An Integrated Approach, Chapter 7.*

Purpose: You will investigate several properties of water including its viscosity, freezing point, boiling point, and surface tension. The solvent properties of water will be examined and you will study saturated, unsaturated, and supersaturated solutions.

Introduction

THE SOLVENT PROPERTIES OF WATER

Water is the most important biological solvent. Blood, urine, perspiration, and lymph are examples of water solutions in the body. The highly polar nature of water allows a large number of solutes to dissolve in it. The rule "like dissolves like" means that most ionic compounds, as well as polar molecular compounds, will dissolve in water. Nonpolar molecular substances, such as gasoline, do not dissolve in water and do not form a solution.

If a substance dissolves in water to form a solution, the mixture will be **homogenous**—it will appear totally uniform with no visible boundaries between one substance and another. If a substance does not dissolve in water, the mixture will be **heterogeneous**—there will be a boundary between the two phases. A cloudy mixture contains tiny pieces of undissolved, suspended solid and is not a solution.

SATURATED, UNSATURATED, AND SUPERSATURATED SOLUTIONS

A **saturated solution** is one in which the solvent has dissolved all the solute it can at a particular temperature. In a saturated solution, dissolved solute and undissolved solute are in equilibrium. The rate at which solid solute dissolves equals the rate at which dissolved solute crystallizes.

If you add many tablespoons of sugar (with stirring) to a cup of water, eventually there will be solid, undissolved sugar on the bottom of the cup. The dissolved sugar is in equilibrium with the undissolved sugar; the solution is saturated with sugar.

$$\text{sugar}_{\text{(undissolved)}} \rightleftharpoons \text{sugar}_{\text{(dissolved)}}$$

If there is only a small amount of solid, undissolved sugar, heating the solution will dissolve the remaining solid. You could stir in a little more sugar and it would dissolve in the hot, unsaturated solution. When sufficient solid sugar has been added, a new saturated solution will be formed. Most liquid and solid solutes become more soluble with increased temperature. A solution may be saturated at one temperature and unsaturated at an elevated temperature.

An **unsaturated solution** contains less solute than the solvent can dissolve at a given temperature. If a little more solute were added, it would dissolve. In an unsaturated solution, dissolved solute and undissolved solute are not in equilibrium—the rate of dissolving is greater than the rate of crystallization.

A **supersaturated solution** is an unstable solution in which the solvent has dissolved more solute than it theoretically can at that temperature. The molecules of a liquid solution are disordered and unorganized; the molecules of a solid crystal are highly organized. To crystallize, a supersaturated solution needs a point of molecular orderliness around which the disordered molecules can crystallize. A disturbance, such as sudden shaking or addition of a tiny piece of solid solute, can furnish the molecular order and cause some of the undissolved solute to form solid crystals. When it crystallizes in this way, the unstable supersaturated solution loses its excess solute and becomes a saturated solution. Honey is an example of a supersaturated sugar solution. Honey sometimes crystallizes as the supersaturated solution turns to a more stable saturated solution. To reform the supersaturated solution, you must heat the solid honey until all solid dissolves, then cool slowly. Most substances crystallize when a saturated solution is cooled, but a few, like honey, cool without crystallizing and become supersaturated.

Supersaturated solutions cannot be formed by stirring additional solute into a saturated solution—the added solute will not dissolve. A supersaturated solution can be made by carefully cooling a hot, saturated solution.

One method to determine whether a solution is saturated, unsaturated, or supersaturated is to add a tiny piece of solid solute. One of three fates awaits the seed crystal—it can dissolve in an unsaturated solution and get smaller, it can stay the same size because a saturated solution cannot dissolve any more solute, or it can get larger because it triggered the crystallization of a supersaturated solution.

Original Solution	*Added Solute Crystal*	*Resultant Solution*
Unsaturated	Gets smaller (dissolves)	Unsaturated or saturated depending on how much solute was added
Saturated	Stays the same size	Saturated
Supersaturated	Gets larger	Saturated

In this experiment you will investigate the solutions of two solutes, sodium chloride and sodium thiosulfate. Sodium chloride is, of course, common table salt. Sodium thiosulfate ($Na_2S_2O_3$) is also known as "hypo" and is used as a fixer in photography. Sodium thiosulfate removes chlorine from solutions in the bleaching of paper pulp. In medicine, it is an antidote for cyanide poisoning.

SURFACE TENSION

Have you ever seen "water bugs" walking across the surface of water in a pond? Water has a kind of "skin" on its surface that holds up the bug. The water molecules on the surface resist being pushed apart from each other by the bug's legs—this resistance to expansion of the surface of a liquid is called **surface tension**. Surface tension is caused by unbalanced forces of attraction. Through hydrogen bonding, water molecules at the surface are pulled inward toward interior water molecules. The force is unbalanced because there are no attractive forces pulling outward on the surface molecules. As a result, any liquid surface becomes as small as possible—this causes the spherical shape of water drops and soap bubbles.

All liquids have surface tension. Water has an unusually high surface tension because its hydrogen bonding is stronger than other types of intermolecular attraction. Detergents weaken the surface tension of water when the detergent molecules move between the water molecules and weaken their attraction for one another.

VISCOSITY

Viscosity is a measure of how easily a liquid flows. High-viscosity liquids like honey resist flow; low-viscosity liquids flow easily. The viscosity of a liquid is affected by changes in temperature—you will investigate the direction of that change in this experiment. Motor oils are available in different viscosities. The viscosity of multi-viscosity motor oils, such as 10W-30, is less affected by temperature changes than single-viscosity oils (such as straight 30 grade viscosity).

You will study the viscosity of several common liquids. Isopropyl alcohol, also known as rubbing alcohol, is a common solvent and is medically useful as an antiseptic. Ethylene glycol is the main ingredient in automotive antifreeze; it is also used in hydraulic brake fluids and as a solvent in ball-point pen ink. Glycerol, also called glycerin, is a moisturizer found in many foods and cosmetics. Glycerol is an important part of triglycerides (fats). Nitroglycerin, the explosive in dynamite, is made from glycerin.

THE ICE POINT

The **freezing point** of a substance is the temperature at which the solid and liquid are in equilibrium. The freezing point of a liquid is the same temperature as the **melting point** of the solid. The **ice point** is the temperature of an ice-water mixture and is approximately the same as the freezing point, but is more easily measured.

You will measure the ice point of both ice-water and of a salt-water solution prepared by adding rock salt to ice-water. Rock salt is naturally occurring, unpurified sodium chloride. Rock salt is used on streets to melt ice. An even better use of rock salt is on ice in making home-made ice cream. In making ice cream we want the ice to absorb heat from the cream so that the cream will freeze. Making ice melt and absorbing heat are different viewpoints of the same phenomenon. The melting of ice is an equilibrium process—ice is melting and water is freezing at the same rate. When ice melts, heat is absorbed from the surroundings; when water freezes, heat is released to the surroundings.

$$\text{ice (at 0°C)} + \text{heat} \rightleftharpoons \text{liquid water (at 0°C)}$$

The **heat of fusion** is the amount of heat required to change one gram of ice (or another solid) at its melting point from a solid to a liquid. Compared to other substances, water has a very high heat of fusion.

When salt, or any other solute, is added to water, the freezing point of the solution is lowered—this is called **freezing point depression**. With added salt, the ice is melting faster than water is freezing; the faster melting ice absorbs heat from its surroundings. Ice on the streets is melted because the salt water solution no longer freezes at 0°C but at a lower temperature—the exact temperature depends on the concentration of the salt solution. A mixture of 33 grams of salt and 100 grams of water will stay liquid until the temperature drops to –21°C. Calcium chloride, $CaCl_2$, lowers the freezing point of water more than sodium chloride, NaCl, because it is more soluble in water (and because each of the three ions, Ca^{2+}, Cl^{1-}, and Cl^{1-}, in the formula unit contributes to the freezing point depression).

THE BOILING POINT

The **boiling point** of a liquid is the temperature at which the vapor pressure of the liquid equals the atmospheric pressure. Since atmospheric pressure varies from location to location and from day to day, the boiling point will change as the pressure varies. The **normal boiling point** of a liquid is the temperature at which the vapor pressure of the liquid equals 1 atmosphere. Water has a normal boiling point of 100°C. At atmospheric pressures less than 1 atmosphere, water will boil at a lower temperature; likewise, at a higher atmospheric pressure it will boil at a temperature greater than 100°C. Pressure cookers cook rapidly because the boiling water is hotter than 100°C; the pressure in the pot has to reach a higher value to become equal to atmospheric pressure.

At temperatures below the boiling point, evaporation occurs only from the surface of the liquid. During boiling, vapor forms throughout the liquid; as the vapor bubbles rise through the liquid, they create the turbulence associated with boiling. Heat must be constantly added to maintain boiling. The heat (**heat of vaporization**) changes the physical state of the water but not its temperature.

$$\text{liquid water (at 100°C)} + \text{heat} \rightleftharpoons \text{water vapor (at 100°C)}$$

The boiling point of a solution is higher than the boiling point of the solvent—that is, adding solute to a solvent raises the solvent's boiling point. This is referred to as **boiling point elevation**.

Procedure

I. SATURATED, UNSATURATED, AND SUPERSATURATED SOLUTIONS

A. Saturated Sodium Chloride Solution

1. Place 3 mL of a saturated sodium chloride solution into each of two 10 × 75-mm test tubes (the smallest of the 3 sizes). Label them A and B.

2. Add a few sodium chloride crystals to test tube A, stopper, and shake vigorously. Note if the crystals dissolve.

3. Add a few sugar crystals to test tube B, stopper, and shake vigorously. Observe.

B. Solutions of Sodium Thiosulfate

1. Into a 25 × 150-mm test tube (the largest of the 3 sizes) place 5 g of sodium thiosulfate, $Na_2S_2O_3$. Add 15 drops of water. Note if the solid dissolves.

2. Add 200 mL of water to a 400-mL beaker and prepare a boiling-water bath as shown in Figure 0-20 and 0-21 on page 20. Place the test tube in the hot-water bath. The test tube can be added before the water reaches a boil. Heat the test tube until a clear solution is obtained; all crystals must be dissolved.

3. Add a single, small crystal of sodium thiosulfate to the hot solution. Observe the crystal as it settles to the bottom of the test tube. Wait 1 minute before proceeding with Step 4.

4. Carefully remove the test tube from the bath and allow it to cool for one hour. Do not agitate the test tube. Save the hot-water bath for Parts II and III.

5. After one hour of cooling, add a single, small crystal of sodium thiosulfate to the solution. Observe the crystal as it settles to the bottom of the test tube. Feel the outside of the test tube for temperature changes.

6. Although it may be hard to see, there is a solution in the test tube. Add a single crystal of sodium thiosulfate to the test tube. Observe what happens to the crystal.

7. Heat the test tube again in the boiling-water bath until a clear solution is obtained.

8. Remove the test tube from the hot-water bath and place it in an ice bath. Allow it to cool for 10 minutes. This rapid cooling may be less successful than the earlier slow cooling.

9. After cooling in ice for 10 minutes, add a crystal of sodium thiosulfate. Observe.

10. Disposal: Heat the test tube in boiling water to dissolve the solid. Pour the hot solution into the "Waste Solution Thiosulfate" container or as directed by your instructor.

II. VISCOSITY

1. Heat the hot-water bath from Part IB to boiling.

2. Using another 400-mL beaker, prepare an ice-water bath containing 200 mL of crushed ice and 20-mL of water.

3. Place eight 16 × 150-mm test tubes (the middle size) in a rack, label them as indicated below, and add the following reagents. Observe any differences in their viscosities as you fill the test tubes. **CAUTION: isopropyl alcohol, ethylene glycol, and glycerol are flammable.**

   No. 1A: 5 mL of isopropyl alcohol
   No. 1B: 5 mL of isopropyl alcohol
   No. 2A: 5 mL of ethylene glycol
   No. 2B: 5 mL of ethylene glycol
   No. 3A: 5 mL of glycerol
   No. 3B: 5 mL of glycerol
   No. 4A: 5 mL of water
   No. 4B: 5 mL of water

4. Place the A test tubes in the ice-water and the B test tubes in the boiling-water bath.

5. After approximately three or four minutes remove each No. 1 test tube from its bath and shake gently. Try to pour each liquid. Note any differences.

6. Repeat Step 5 with No. 2, No. 3, and No. 4 test tubes.

7. Save the boiling-water bath for Part III.

III. THE BOILING POINT OF WATER

1. Heat the hot-water bath you used in Part II until the water boils. Make sure the water has come to a full, rolling boil.

2. Insert a thermometer into the boiling water, and when the thermometer displays a constant temperature, record the temperature to the precision allowed by the thermometer.

IV. THE ICE POINT OF WATER AND OF A SOLUTION

1. Use a 250-mL beaker to prepare a new ice bath containing 150 mL of crushed ice and 10 mL of water.

2. Immerse a thermometer in the water-ice slurry and stir **with a stirring rod** for at least two minutes. Leave the thermometer in the ice bath but be careful not to jar it with the stirring rod. When the column in the thermometer stops moving or the reading on a digital thermometer is constant, read the thermometer to the nearest 0.1°C.

3. Continue stirring and repeat the thermometer measurement twice more at two minute intervals.

4. Add 1 heaping tablespoon (about 20 grams) of rock salt, thoroughly stir for two minutes, and take a temperature reading.

5. Continue the thorough stirring and temperature measurement twice more at two minute intervals. Note if the salt melts the ice.

V. SURFACE TENSION

1. Fill a shallow, soap-free metal pan with water to a depth of about 1 or 2 cm.

2. Float a small piece of tissue paper (about 3 cm × 3 cm) on the surface of the water. Rest a soap-free straight pin on the tissue paper.

3. Slowly squirt liquid dish soap into the water on the opposite side of the pan from the pin. Observe the behavior of the pin.

VI. SOLUBILITIES OF COMMON SUBSTANCES IN WATER

1. Place seven 16 × 150-mm test tubes (the middle size) in a rack, number them, and add the following substances:
   - No. 1: A small amount on the end of a spatula, about the size of a pea of table salt (sodium chloride, NaCl)
   - No. 2: A small amount on the end of a spatula, about the size of a pea of table sugar (sucrose, $C_{12}H_{22}O_{11}$)
   - No. 3: 1 mL of cooking oil (vegetable oil, a mixture, typically $C_{39}H_{66}O_6$)
   - No. 4: A small amount on the end of a spatula, about the size of a pea of barium sulfate ($BaSO_4$, an X-ray medium)
   - No. 5: 1 mL of isopropyl alcohol (rubbing alcohol, C_3H_8O) **CAUTION, flammable.**
   - No. 6: A small amount on the end of a spatula, about the size of a pea of potassium nitrate (saltpeter, KNO_3)
   - No. 7: 1 mL of glacial acetic acid (a component of vinegar, $HC_2H_3O_2$). **CAUTION, strong odor and corrosive.**

2. Add 5 mL of deionized water to each of the test tubes, stopper them, and mix for a minute or two. If the mixture is homogeneous (if there is only one phase), the substance is soluble in water. If there is more than one phase, the substance is insoluble in water. Use S or I on the Report Sheet to indicate solubility or insolubility.

3. Disposal: Place the cooking oil (No. 3) in the "Waste Oil" container. Place the barium sulfate (No. 4) in the "Waste Barium Sulfate" container. The other mixtures may be disposed in the sink or as directed by your instructor. Use detergent and a brush to clean the test tubes.

Name ______________________ Section ____________ Date ____________

Prelab Questions for Experiment 8

1. When a sodium chloride crystal is dropped into a solution, it gets smaller in size and disappears. Was the solution saturated, unsaturated, or supersaturated with sodium chloride?

   ____

2. When a saturated solution is heated, does it usually remain saturated, become unsaturated, or supersaturated?

   ____

3. When two liquids are mixed together in a test tube, only one phase results. Are the liquids soluble or insoluble in each other?

   ____

4. The following substances are used in Part VI of this experiment. Use I or M to classify each as an ionic compound or a molecular compound. Hint: review Experiment 4, page 75.

   ________ salt, NaCl

   ________ sugar, $C_{12}H_{22}O_{11}$

   ________ cooking oil, $C_{39}H_{66}O_6$

   ________ barium sulfate, $BaSO_4$

   ________ isopropyl alcohol, C_3H_8O

   ________ potassium nitrate, KNO_3

   ________ acetic acid, $HC_2H_3O_2$

5. Are ionic compounds usually soluble or insoluble in water?

   ____

6. If a molecular compound dissolves in water, is it a polar molecular compound or a nonpolar molecular compound?

   ____

Name ______________________ Section ____________ Date ____________

Report for Experiment 8

I. SATURATED, UNSATURATED, AND SUPERSATURATED SOLUTIONS

A. Saturated Sodium Chloride Solution

1. a. Did the sodium chloride crystals dissolve when you added them to test tube A?

   ____

   b. Explain your observations.

   ____

   ____

2. a. Did the sugar crystals dissolve as you added them to test tube B? ____________

   b. Explain your observations.

   ____

   ____

B. Solutions of Sodium Thiosulfate

1. a. What happened in Step 3 when you added a crystal of sodium thiosulfate to the hot sodium thiosulfate solution?

   ____

   b. Before you added the crystal in Step 3, was the hot sodium thiosulfate solution saturated, unsaturated, or supersaturated?

   ____

   c. After you added the crystal in Step 3, was the hot sodium thiosulfate solution saturated, unsaturated, or supersaturated?

   ____

2. a. What happened in Step 5 when you added a crystal of sodium thiosulfate to the cold sodium thiosulfate solution?

   ____

b. Before you added the crystal in Step 5, was the cold sodium thiosulfate solution saturated, unsaturated, or supersaturated?

____

c. After you added the crystal in Step 5, was the cold sodium thiosulfate solution saturated, unsaturated, or supersaturated?

____

d. Is the crystallization of sodium thiosulfate an exothermic or an endothermic process?

____

3. a. What happened in Step 6 when you added a crystal of sodium thiosulfate to the hard-to-see sodium thiosulfate solution?

____

b. Before you added the crystal in Step 6, was the hard-to-see sodium thiosulfate solution saturated, unsaturated, or supersaturated?

____

c. After you added the crystal in Step 6, was the hard-to-see sodium thiosulfate solution saturated, unsaturated, or supersaturated?

____

4. Compare the results of adding a sodium thiosulfate crystal to the slowly cooled solution in Step 9 to results of adding a sodium thiosulfate crystal to the rapidly cooled solution in Step 5. Did the solutions behave in the same way?

____

II. VISCOSITY

1. Which of the four reagents (isopropyl alcohol, ethylene glycol, glycerol, or water) is most viscous at room temperature?

____

2. How was the viscosity of the four reagents affected by lowering the temperature?

____

3. How was the viscosity of the four reagents affected by increasing temperature?

____

Name ______________________________ Section ____________ Date ____________

III. THE BOILING POINT OF WATER

1. Data. Observed boiling point of water: ______________

2. Explain why your measured boiling point might be different from the accepted boiling point of water.

____

____

IV. THE ICE POINT OF WATER AND OF A SOLUTION

1. a. Data. Ice Point of Water:

	Trial 1	*Trial 2*	*Trial 3*	*Average*
Ice point of water				

b. In this procedure, you measured the temperature of the ice-water mixture every two minutes as it sat at room temperature. If you continued stirring the ice-water mixture, when would the temperature of the ice-water increase?

____

2. a. Data. Ice Point of Saltwater:

	Trial 1	*Trial 2*	*Trial 3*	*Average*
Ice point of saltwater				

b. Explain why the ice point of saltwater is different from the ice point of water.

____

____

V. SURFACE TENSION

1. Why did the straight pin seem to float on the surface of the water?

____

2. What happened to the pin as you added the soap?

____

3. Explain why the soap affected the pin in this manner.

____

VI. SOLUBILITIES OF COMMON SUBSTANCES IN WATER

1. Data:

	Solute	*Water Solubility (S or I)*
1	sodium chloride, NaCl	
2	sugar, $C_{12}H_{22}O_{11}$	
3	cooking oil, $C_{39}H_{66}O_6$	
4	barium sulfate, $BaSO_4$	
5	isopropyl alcohol, C_3H_8O	
6	potassium nitrate, KNO_3	
7	acetic acid, $HC_2H_3O_2$	

2. Ionic compounds are usually water soluble. List the soluble and the insoluble ionic compounds you tested.

Water-soluble Ionic Compounds	*Insoluble Ionic Compounds*

3. Molecular compounds that dissolve in water are polar; insoluble molecular compounds are nonpolar. Classify the molecular compounds tested as polar or nonpolar.

Polar Molecular Compounds	*Nonpolar Molecular Compounds*

Name ______________________ Section ____________ Date ______________

Related Questions for Experiment 8

1. When classifying a solution as saturated or unsaturated, why is it important to specify the temperature?

____

____

2. What usually happens to the viscosity of a liquid when it is heated?

____

3. Explain the difference in the boiling points of water (100°C) and seawater (102°C).

____

____

4. Explain the difference in the freezing points of water (0°C) and seawater (–8°C).

____

____

5. Explain why the boiling point of water averages 94°C in Denver, Colorado, which is 1.6 km (1 mile) above sea level.

____

____

6. Pure ethylene glycol (antifreeze) has a freezing point of –11.5°C and water has a freezing point of 0°C. Explain why a 50-50 mixture of ethylene glycol and water remains liquid at –40°C.

____

____

7. Explain how the combined use of an ethylene glycol-water antifreeze solution (rather than only water) and a pressurized system can allow automobile cooling systems to reach temperatures of more than 120°C without boiling.

____

____

8. If the pressurized radiator cap is removed from a hot car radiator, steam and boiling coolant may suddenly escape and cause severe burns. Explain why the coolant suddenly boils when the radiator cap is removed.

____

____

9. Explain why ice melts when rock salt is poured on its surface.

____

____

10. Why is a rock salt-ice mixture used in the preparation of home-made ice cream?

____

____

Experiment 9
Ions of Nutrition

Purpose: You will investigate the presence of several ions, found in products you encounter every day, which are vital to the living organism.

Introduction

Of the 90 elements naturally found on earth, 25 are essential to living organisms. The major elements in all living organisms are hydrogen, oxygen, carbon, and nitrogen—they are found in organic compounds such as carbohydrates, fats, proteins, and vitamins. The remaining 21 elements are **minerals**—these are naturally occurring inorganic elements. The minerals can be divided into two groups, the major minerals and the trace minerals.

The **major minerals**, or macrominerals, are found in greater mass and include calcium, phosphorus, potassium, sulfur, chloride, sodium, and magnesium. Together, calcium and phosphorus make up three-fourths of the mass of all the minerals present in your body. The **trace minerals** are iron, iodide, fluoride, manganese, zinc, molybdenum, copper, cobalt, chromium, selenium, arsenic, nickel, silicon, and boron. Even though the trace minerals are found in lower amounts than the major minerals they are equally important to your body. A daily deficiency of a few micrograms of iodine is just as serious as a deficiency of several hundred milligrams of calcium. Other minerals such as gold and aluminum are also found in the body but are not known to be nutrients.

The procedures in this experiment are **qualitative** tests—they determine *what* substance is present. These same procedures could be modified for **quantitative** analysis—to determine *how much* of the substance is present.

IODIDE ION

The element iodine usually is found in nature as the **iodide ion** (I^-). Iodide ion is needed by the body in an extremely small quantity, but obtaining this amount is critical. The body uses iodide ion to make **thyroxin**, the hormone responsible for regulating the basal metabolic rate. Without iodide ion, the body cannot make thyroxin. Thyroxin is synthesized in the **thyroid**, a gland located in the lower neck. When the iodide level of the blood is low, the cells of the thyroid enlarge, forming a **goiter**. People with this condition suffer sluggishness and weight gain. Infants born to mothers suffering from this condition may be born with irreversible mental and physical retardation known as **cretinism**.

A government appointed committee of the U.S. National Academy of Sciences/National Research Council publishes recommendations concerning appropriate nutrient intakes for people in the United States. These are **recommended daily allowances (RDA)**. RDAs are the suggested average daily consumption levels of energy and selected nutrients. RDAs have been set for protein, eleven vitamins, and seven minerals. RDAs have not been established for all nutrients but the committee has published recommended ranges of safe and adequate intake for two more vitamins, eight more minerals, and energy (Calories). Separate RDAs have been set for different sets of people: men, women, pregnant women, lactating women, children, and different age groups. The U.S. RDA of iodine is 0.15 mg for adult men and women.

Iodide ion is found in some soils and is therefore present in plants growing in that soil. A more plentiful source is the ocean, so seafood is a dependable source of iodine. To prevent dietary shortages of iodine, iodide ion (in the form of potassium iodide) can be added to table salt. The salt-box label will state whether the salt is iodized. Iodine is also added to milk and bakery products.

Excessive intake of iodine can also enlarge the thyroid. The level at which detectable harm results is only a few times higher than most people receive daily. Like other nutrients, iodide ion is a deadly poison in large amounts. The RDAs attempt to ensure adequate dietary nutrition but they do not warn against overnutrition. Separate recommendations are made for maximum daily intake of nutrients.

In this experiment, salt will be tested for the presence of iodide ion. Iodide ion (I^-) is oxidized to free iodine (I_2) by chlorine-water, a mixture of C_{12} and water.

$$2\,I^- + Cl_2 \longrightarrow 2\,Cl^- + I_2$$

Iodide ion is very soluble in water; iodine is only slightly soluble in water. Iodine is much more soluble in less polar solvents such as dichloromethane. Dichloromethane and water are **immiscible**—they do not mix. When iodine dissolves in dichloromethane, the solution turns a violet color.

CALCIUM ION

Calcium is essential to bone formation, tooth formation, nerve transmission, maintenance of normal blood pressure, blood clotting, muscle contraction, and heart function. Cells need continuous access to calcium, so the body maintains a calcium ion concentration in the blood. The skeleton serves as a bank from which the blood can borrow and return calcium. You can go without adequate calcium for years without suffering noticeable symptoms and then, late in life discover that your calcium savings are depleted and the integrity of your skeleton can no longer be maintained. Adult bone loss, or **osteoporosis**, is a health problem for many older people. Each year over a million people in the U.S. suffer bone breaks due to osteoporosis. Menopause increases bone loss in women.

The recommended daily allowance of calcium varies by age, ranging between 400 mg for infants and 1200 mg for ages 11-24; the RDA of calcium is also 1200 mg for pregnant and lactating women. After age 24 the RDA is set at 800 mg. The RDA for young people is high because people develop their peak bone mass during this period. Unfortunately, few girls and women meet the RDA during their bone-forming years. This may mean that they start their adult lives with less than optimal bone density.

Calcium is not widespread in the diet. Milk and milk products are the most important dietary sources of calcium. People who are unwilling or unable to consume milk products or other calcium-rich foods may choose calcium supplements instead. Regular vitamin-mineral pills do not contain a significant amount of calcium. One problem concerning the use of calcium supplements is possible excessive calcium intake which can cause urinary tract stones, kidney damage, vitamin D toxicity, low iron levels, drug interactions, constipation and confusion.

In this experiment, milk will be tested for calcium. When ammonium oxalate, $(NH_4)_2C_2O_4$, is added to calcium ion, a precipitate of calcium oxalate, CaC_2O_4, is formed.

$$(NH_4)_2C_2O_4 + Ca^{2+} \longrightarrow CaC_2O_4 + 2\ NH_4^+$$

PHOSPHATE ION

The element phosphorus is needed primarily as **phosphate ion**, PO_4^{3-}. Together with calcium ion, phosphate ion is essential to bone and tooth formation. Phosphates are important constituents of **buffers** in the blood which maintain blood acid-base balance. Phosphorus compounds are essential to energy transfer in cells via adenosine triphosphate (ATP). Phosphorus is an essential part of the genetic materials DNA and RNA. Certain lipids—**phospholipids**—that form the membranes around each cell also contain phosphorus.

Phosphorus needs are easily met by most any diet and deficiencies are unknown. The U.S. RDA for phosphorus is 1200 mg for ages 11-24, and for pregnant and lactating women. After age 24 the RDA is 800 mg. Excess phosphorus intake may cause calcium excretion.

In this experiment, milk will be tested for phosphate ion. Phosphate ion may be detected by treating the milk with a few drops of nitric acid and ammonium molybdate. The precipitate formed, ammonium phosphomolybdate, confirms the presence of phosphate ion.

IRON ION

Iron is primarily needed as a component of **hemoglobin** and **myoglobin**. Both compounds use iron to carry or hold oxygen. Hemoglobin is the oxygen carrier in the blood and myoglobin is the oxygen reservoir in the muscle cells. Iron is also used by many enzymes in the metabolic transfer of energy.

Iron-deficiency anemia is a condition where the red blood cells contain too little iron and thus deliver too little oxygen to the tissues. Symptoms include tiredness, apathy, inability to concentrate, and a tendency to feel cold. In the U.S., some eight percent of women and one percent of men have anemia; many more have only borderline reservoirs of iron. Worldwide, the incidence of iron deficiency is more pronounced—over a billion people suffer from anemia.

Iron deficiency is usually caused by sheer lack of food or high consumption of the wrong foods. In the U.S. the cause is too few iron-rich foods and too many foods high in sugar and fat. Iron deficiency may also be caused by poor absorption of the iron in sugar and fat. Iron deficiency may also be caused by poor absorption of the iron present in food. Tea, coffee, wheat bran, fiber, and calcium supplements decrease the absorption of iron. Even a normal person absorbs only about 10 percent of the iron in the diet. Iron in meat, fish, and poultry is more easily absorbed than that in other sources such as egg yolks, legumes, and dried fruits. Vitamin C enhances the absorption of iron.

The iron U.S. RDA is 10 mg for men and older women. For women of childbearing age the RDA is 15 mg; pregnant women should receive 30 mg of iron daily. 80 percent of the body's iron is in the blood, so loss of blood causes loss of iron. Because of menstruation, a woman's need for iron is greater than a man's.

The U.S. diet usually does not provide enough iron. It is best to rely on foods to meet this need rather than iron supplements; however, iron supplements are usually recommended for pregnant women. About 25 percent of your iron comes from "enriched" or "fortified" foods that have iron added during processing. These include enriched bread and fortified breakfast cereals. Cooking in iron pans adds substantial amounts of iron to your food just as cooking in aluminum pans adds aluminum to your food. Aluminum is not known to be a nutrient and some have suggested that dietary aluminum may be harmful.

Iron is toxic in large amounts, and once absorbed it is difficult to excrete. Even for healthy men, the safety of high levels of iron intake is questionable. Excess iron may cause liver injury, a greater risk of infections, and possibly increased incidence of heart disease.

Iron can be found in three forms. **Elemental iron** (Fe) is the metallic solid used to make iron pans. Elemental iron atoms are neutral; they have no charge. Tiny iron metal filings are sometimes added to foods. Elemental iron may be listed on an ingredients label as "iron" or as "reduced iron." In the stomach, elemental iron is slowly oxidized* by hydrochloric acid to iron(II) ion.

The second form of iron is iron(II) ion, also called **ferrous** ion. Iron(II) ion has a positive two charge, Fe^{2+}. Iron(III) ion or **ferric** ion, Fe^{3+}, is the third form of iron. Ferrous iron is more easily absorbed than ferric iron. Ferrous iron is somewhat unstable; it is easily oxidized to ferric ion by the oxygen in the air.

In this experiment, iron in breakfast cereal will be detected in two ways. Elemental iron is attracted to a magnet; a Teflon-covered magnetic stirring bar can be used to extract the elemental iron from a breakfast cereal.

Iron(II) ion is not attracted to a magnet but can be detected by its reaction with potassium ferricyanide, $K_3Fe(CN)_6$ to form a dark blue precipitate. Iron(III) reacts with potassium ferricyanide to form a green precipitate.

$$3\,Fe^{2+} + 2\,Fe(CN)_6^{3-} \longrightarrow \underset{\text{(dark blue)}}{Fe_3[Fe(CN)_6]_2}$$

$$Fe^{3+} + Fe(CN)_6^{3-} \longrightarrow \underset{\text{(green)}}{Fe[Fe(CN)_6]}$$

*Elemental iron has an **oxidation number** of zero. The oxidation number of Fe^{2+} is +2 and the oxidation number of Fe^{3+} is +3. An **oxidation** is a chemical reaction that increases the oxidation number of a substance. During oxidation a substance loses one or more electrons. A **reduction** is a chemical reaction that decreases the oxidation number of a substance. During reduction a substance gains one or more electrons. Oxidation and reduction must occur simultaneously; while one substance is oxidized another is reduced.

PROCEDURE

I. DETECTION OF IODINE IN IODIZED SALT

WE RECOMMEND THAT STEP 1 OF THIS EXPERIMENT IS CONDUCTED BY AN INSTRUCTOR, DUE TO SAFETY ISSUES AND THE USE OF IODINE

1. Dissolve a crystal of iodine, no larger than a grain of ice, in a regular (16 × 125 mm) test tube containing 4 mL of deionized water. Note the color of the solution.

2. Add 1 mL of dichloromethane, stopper the test tube, and shake. Note the color of the two layers.

3. Use a mortar and pestle to pulverize approximately 10 g of iodized salt. Transfer the salt to a 150-mL beaker.

4. Add 15 mL of 95% ethanol and place the beaker on a magnetic stirrer. Add a magnetic stirring bar and stir the suspension for approximately 30 minutes. Most of the potassium iodide will dissolve in the ethanol; the sodium chloride will not. (If a magnetic stirrer is not available, stir the solution by hand periodically for about 45 minutes.)

5. Repeat steps 3 and 4 with noniodized salt. Proceed to Part II of this experiment while waiting for the potassium iodide to dissolve.

Follow steps 6 through 10 with each of the two salt suspensions.

6. Filter the suspension. Collect the filtrate in an evaporating dish. (See Fundamental Laboratory Operations, Part 8, page 23.)

7. Evaporate the solution to dryness over a hot water bath using a hotplate. (See Fundamental Laboratory Operations, Part 7b, page 21.)

CAUTION: Ethanol is flammable. This procedure should only be done in a fume hood using a hotplate. If the ethanol should catch on fire, cover the evaporating dish with a watch glass using a pair of tongs.

8. Allow the evaporating dish to cool.

9. Dissolve the residue of the evaporating dish in about 5 mL of deionized water and transfer the solution to a 16 × 125-mm test tube.

10. Add 1 mL of dichloromethane and 2 mL of freshly prepared chlorine-water to the test tube.

11. Stopper the test tube and shake. Note the color of the two layers.

12. DISPOSAL: Pour the dichloromethane solutions into the "Waste Halogenated Solvent" container or as directed by your instructor.

II. DETECTION OF CALCIUM ION IN MILK

1. Pour 20 mL of whole milk into a 250-mL beaker. Add 20 mL of deionized water.

2. Slowly add, while stirring, 10 mL of a 5% solution of acetic acid. The milk fat and casein protein will precipitate.

3. After the precipitate settles, filter the solution. Collect the filtrate in a 250-mL beaker. [If necessary, filter with suction, using a Buchner funnel, filter flask, and trap. Pour the filtrate into a 250-mL beaker.]

4. Pour 2 mL of the filtrate into a test tube and add 2 mL of 4% ammonium oxalate, $(NH_4)_2C_2O_4$. A precipitate (calcium oxalate) indicates the presence of calcium.

5. Save the remainder of the filtrate for use in Parts III and IV.

6. Repeat steps 1 through 5 with skim (nonfat) milk.

7. Dispose of the contents of the test tubes as directed by your instructor.

III. DETECTION OF PHOSPHATE ION IN MILK

Follow steps 1 and 2 with each of the two filtrates (whole milk and skim milk) from Part II.

1. Pour 3 mL of the filtrate into a test tube, then add about 3 drops of concentrated nitric acid (HNO_3) and 2 drops of 2.5% ammonium molybdate solution.

CAUTION: Nitric acid is a strong acid. Wash your hands immediately after handling concentrated nitric acid. Notify your instructor about spills on the bench top.

2. Place the test tube in a hot-water bath using a hotplate. The formation of a precipitate indicates the presence of phosphate ion.

3. Dispose of the contents of the test tubes as directed by your instructor.

IV. DETECTION OF IRON

A. Iron in Milk

1. Pour 5 mL of the whole milk filtrate from Procedure II into a regular (16 × 125 mm) test tube and add 1 mL of 0.2 M $K_3Fe(CN)_6$ (potassium ferricyanide). A dark blue color indicates the presence of iron.

2. Repeat Step 1 with skim milk filtrate.

3. Dispose of the contents of the test tubes as directed by your instructor.

B. Elemental Iron in Breakfast Cereal

1. Place a single serving of an instant hot breakfast cereal in a plastic sandwich-size zip-close bag. Add a Teflon-coated magnetic stirring bar to the bag. Securely close the bag.

2. Shake the bag for 2 or 3 minutes.

3. Remove the stirring bar and look for evidence of iron. The iron filings are very tiny. It may help to hold the stirring bar under a gentle stream of water; the water will wash the iron to the end of the stirring bar.

4. Clean the stirring bar by wiping the iron onto a paper towel or by washing off the iron with a strong stream of water.

C. Iron Ions in Breakfast Cereal

It is recommended that Steps 1 to 3 be done in a fume hood. Wear gloves to handle the acid and the filtrate.

1. Weigh approximately 10 g of breakfast cereal in a 150-mL beaker, and add 25 mL of 6M HCl (hydrochloric acid). (**CAUTION!**)

2. Cover the beaker with a small watch glass, and place it for about 15 minutes in a 400-mL beaker half-filled with warm water (70°C).

3. Remove the beaker from the water bath, allow the solution to cool, and filter it. Collect the filtrate in a small beaker.

4. Pour 5 mL of the filtrate into a test tube, and add 1 mL of 0.1 M $K_3Fe(CN)_6$ (potassium ferricyanide). Allow the solution to stand for at least 10 minutes. A dark blue color indicates the presence of iron.

5. Repeat Steps 1 through 4 with any other cereal assigned to you by your instructor. Compare the colors produced in Step 4.

6. Dispose of the contents of the test tubes as directed by your instructor.

Name ______________________ Section ____________ Date ______________

Prelab Questions for Experiment 9

________ 1. What substance turns dichloromethane to a violet color?
- a. iodide ion
- b. iodine
- c. calcium ion
- d. phosphate ion

________ 2. In Part I, the purpose of adding chlorine water is to
- a. make iodine more soluble.
- b. change iodide ion to iodine.
- c. change iodine to iodide ion.
- d. make dichloromethane immiscible with water.

________ 3. In Part II, the purpose of adding acetic acid is to
- a. precipitate calcium ion.
- b. remove iodide ion.
- c. precipitate milk fat and casein.
- d. precipitate albumins and globulins.

________ 4. What substance forms a precipitate with ammonium oxalate?
- a. iodide ion
- b. iodine
- c. calcium ion
- d. phosphate ion

________ 5. What substance forms a precipitate with ammonium molybdate?
- a. iodide ion
- b. iodine
- c. calcium ion
- d. phosphate ion

________ 6. What substance reacts with potassium ferricyanide to form a dark blue product?
- a. Fe^{2+} ion
- b. Fe^{3+} ion
- c. both Fe^{2+} and Fe^{3+} ions
- d. Fe

7. What does RDA mean?

____

____

8. Some people take dietary supplements that provide huge doses of vitamins and minerals equal to many times the RDA. Explain why this may be unhealthy.

____

____

9. Draw the electron dot diagram of iodide ion and iodine.

   a. Iodide ion, I^-

   b. Iodine molecule, I_2

Name ______________________ Section ____________ Date ______________

Report for Experiment 9

I. DETECTION OF IODINE IN IODIZED SALT

1. a. How were you able to distinguish the dichloromethane from the water layer?

   b. Is dichloromethane more dense or less dense than water?

2. Record the color of the following solutions.

	Water	*Dichloromethane*
Iodine (Steps 1 and 2)		
Iodized salt (Step 10)		
Noniodized salt (Step 10)		

3. Explain how this procedure confirms the presence of iodine in iodized salt.

____

____

II. DETECTION OF CALCIUM IN MILK

1. a. Is there a significant difference between the fat content of whole milk and skim milk?

   ____

   b. Explain.

   ____

2. a. Is there a significant difference between the calcium content of whole milk and skim milk?

   ____

   b. Explain.

   ____

III. DETECTION OF PHOSPHATE ION IN MILK

1. a. Is there a significant difference between the phosphate content of whole milk and skim milk?

   ____

   b. Explain

   ____

IV. DETECTION OF IRON

A. Iron in Milk

Use a (+) for a positive and a (–) for a negative test for iron.

	Results of $K_3Fe(CN)_6$ test
Whole Milk	
Skim Milk	

B. Elemental Iron in Breakfast Cereal

Use a (+) for a positive and a (–) for a negative test for iron.

Name of Cereal	*Iron Filings on Magnet*

C. Iron Ions in Breakfast Cereal

Use a (+) for a positive and a (–) for a negative test for iron.

Name of Cereal	*Results of $K_3Fe(CN)_6$ test*

Name ______________________ Section __________ Date ____________

Related Questions for Experiment 9

1. Dietary sodium occurs as sodium ion, Na^+. Diets rarely lack sodium and no RDA has been set for sodium. The estimated minimum daily sodium requirement is 500 mg. Excessive intake of sodium has been linked to high blood pressure and heart disease. The recommended maximum daily intake is 2400 mg sodium, the equivalent of about 1 teaspoon of salt. Americans generally consume at least twice that amount. About 10 percent of your sodium intake is due to sodium that occurs naturally in foods. Your kitchen saltshaker may account for as little as 15 percent of your total sodium intake. Up to 75 percent may be due to processed foods. For example, a one half-cup serving of canned chicken noodle soup (less than half a can) contains 950 mg of sodium. Some frequently taken over-the-counter medicines contain surprisingly large amounts of sodium; one dose of Original Alka-Seltzer® with aspirin contains more than 100 mg of sodium.

   Calculate the recommended daily maximum sodium intake in grams.

2. Write the formula (including the charge) of the following ions of nutrition.

   a. calcium ion

   b. chloride ion

   c. magnesium ion

   d. phosphate ion

   e. sodium ion

   f. iron(II) ion (ferric ion)

   g. iron(III) ion (ferric ion)

   h. fluoride ion

   i. iodide ion

   j. sulfate ion

3. Write the formula of the following compounds found as food additives or food preservatives.

   a. potassium iodide

   b. sodium bicarbonate

   c. sulfur dioxide

   d. calcium carbonate

   e. sodium chloride

   f. ferrous sulfate

   g. silicon dioxide

   h. ammonium sulfate

   i. magnesium carbonate

   j. ferric phosphate

   k. sodium sulfite

   l. sodium bisulfite

   m. potassium chloride

   n. magnesium oxide

   o. sodium carbonate

   p. calcium sulfate

   q. sodium phosphate

Experiment 10
Lipids

Reference: *General, Organic, and Biological Chemistry: An Integrated Approach, Chapter 8.*

Purpose: You will investigate several physical and chemical properties of lipids including solubility, identification tests, and tests for rancidity and unsaturation. You will prepare soap and compare its action to a detergent in hard and soft water.

Introduction

Lipids are oily or waxy substances that may be extracted from tissues using nonpolar solvents. They are grouped into two major classes: those that can be saponified (hydrolyzed by a base) and those that cannot be saponified.

SAPONIFIABLE LIPIDS

The saponifiable lipids you will study in this experiment are **triglycerides***—triesters of glycerol and three fatty acids. About 95 percent of the lipids in foods and in human bodies are triglycerides. Triglycerides from plants tend to be liquids and are called **oils**; **fats** are solid triglycerides from animals. **Saturated fatty acids** have only single bonds between carbon atoms. **Unsaturated fatty acids** contain carbon-to-carbon double bonds. Unsaturated fatty acids with one carbon-carbon double bond are **monounsaturated**; those with two or more carbon-carbon double bonds are **polyunsaturated**. The fatty acids in animal fat tend to be saturated fatty acids; examples include the "marbling" in beef, chicken fat, butter, and lard. Fatty acids in vegetable oils are generally higher in unsaturated fat but vary considerably from vegetable to vegetable. Tropical oils such as coconut oil and palm oil are high in saturated fats. Olive oil and canola (rapeseed) oil are high in monounsaturated fatty acids while safflower oil, sunflower oil, and corn oil are high in polyunsaturated fats. In this experiment you will investigate cottonseed oil (about 25% saturated, 25% monounsaturated, and 50% polyunsaturated fatty acids) as well as commercial vegetable oils.

Much remains unknown about the health effects of dietary fat but current research indicates that Americans consume far too much fat. The amounts of triglycerides and cholesterol found in the blood are useful in assessing a person's risk of cardiovascular (heart and artery) disease. Most physicians recommend reducing the amount of fat in the diet. The recommendation to substitute unsaturated fats for saturated fats has recently been revised since research has shown that unsaturated trans fatty acids also play a role in increasing the level of cholesterol in the blood. People's love for fat has prompted great interest in artificial fats which are just coming to the market.

---

*Triglycerides are more accurately named **triacylglycerols** but the term triglyceride is still commonly used.

Margarine and solid shortening (such as Crisco®) are made by hydrogenation of unsaturated vegetable oils. The addition of hydrogen to carbon-carbon double bonds raises the melting point of the triglyceride; where vegetable oils are liquid, hydrogenated vegetable oils are solid. Natural vegetable oils contain only *cis*- double bonds. Partially hydrogenated margarine and shortening contain up to 40 percent of the *trans*- isomer. Recent research suggests the *trans*- isomers of margarine and shortening are linked to lower HDL (High-Density Lipids) levels and may increase the rate of heart disease.

People need *some* fat in their diet; linoleic acid and linolenic acid are essential fatty acids that human beings cannot synthesize. These fatty acids or their derivatives help regulate the formation of blood clots, blood pressure, the immune response, the inflammation response to injury and many other body functions. Linoleic acid and linolenic acid are polyunsaturated fatty acids; both are widely distributed in foods and are easily stored in the body, so deficiencies are unlikely.

The degree of relative unsaturation of fats and oils can be compared by measuring the quantity of bromine that will react with equal amounts of lipids. Recall the test for unsaturation performed in Experiment 5:

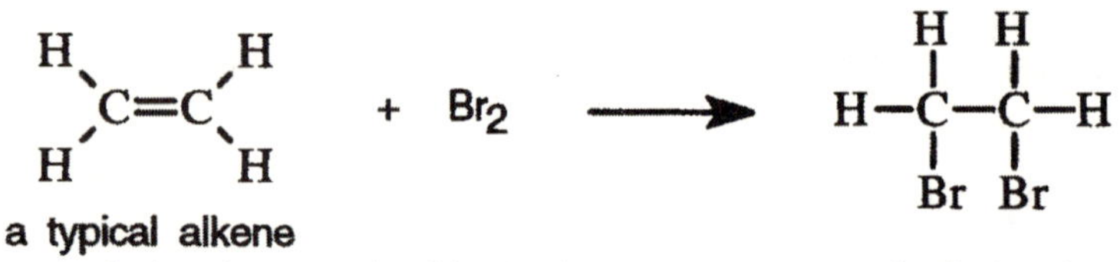

Oils are more easily digested than fats. They become rancid more easily than fats because of the ease in which their double bonds can be hydrolyzed or oxidized. The most commonly used test to determine oxidative rancidity is the Kreis test. A small quantity of concentrated hydrochloric acid is added to the lipid to be tested, followed by the addition of a 0.1% solution of phloroglucinol (1,3,5-trihydroxybenzene) in ether. A red or pink color indicates rancidity.

Hydrolysis (splitting apart by the addition of water) of a fat or oil produces glycerol and three fatty acids.

$$\text{fat or oil} + 3\ H_2O \longrightarrow \text{glycerol} + 3\ \text{fatty acids}$$

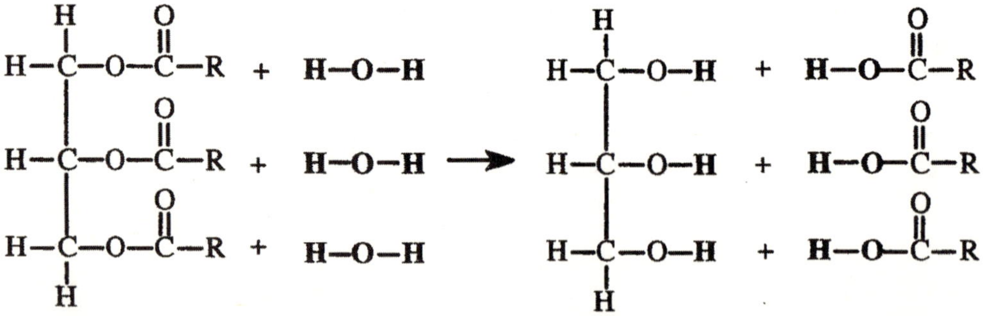

When hydrolysis occurs in the presence of a base, soap and glycerol are formed. Soap is the salt of a fatty acid; hydrolysis by a strong base is called **saponification**. In Experiment 15 you will saponify the ester group of aspirin. In the experiment you will saponify a lipid.

fat or oil + 3 NaOH ⟶ glycerol + soap

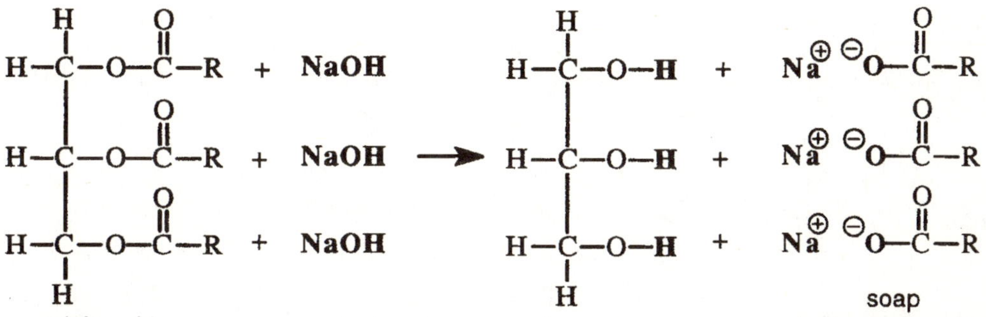

triglyceride sodium hydroxide glycerol soap (salts of fatty acids)

When heated with potassium hydrogensulfate, $KHSO_4$, the glycerol is dehydrated and oxidized to acrolein, which may be detected by its sharp irritating odor. The distinctive odor of acrolein thus serves as the basis of the test for a saponifiable lipid.

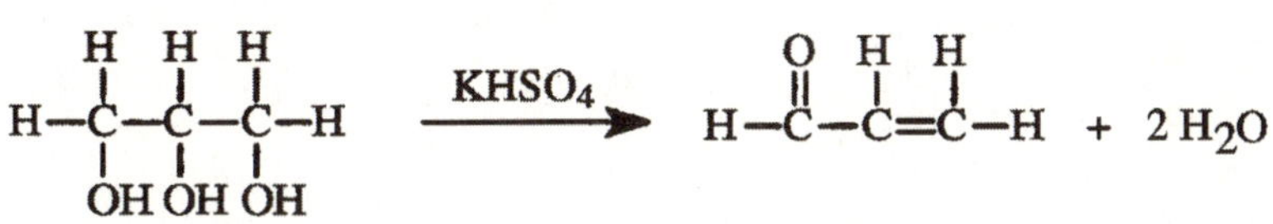

glycerol acrolein

SOAPS AND DETERGENTS

A **soap** is the sodium or potassium salt of a fatty acid. The greater part of the soap molecule is nonpolar, but the polar "head" at the carboxyl end permits a small portion of the molecule to be water soluble. The cleansing power of soap is dependent on this water solubility. Water containing Ca^{2+} and Mg^{2+} ("hard" water) forms insoluble salts with soap and reduces its cleansing properties. Water softeners have ion-exchange resins that replace Ca^{2+} and Mg^{2+} with Na^+ which does not interfere with the action of soap.

A **detergent** is the sodium salt of a sulfonic acid. It has similar cleansing properties to soap but has many significant advantages. Calcium and magnesium ions have no effect on the cleansing properties of a detergent, since their calcium and magnesium salts are water soluble.

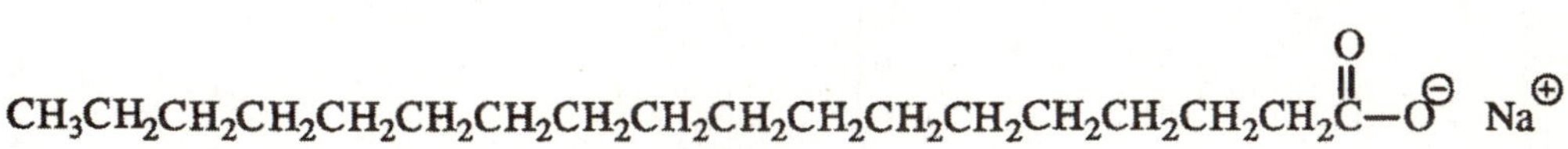

a soap
(the sodium salt of a fatty acid)

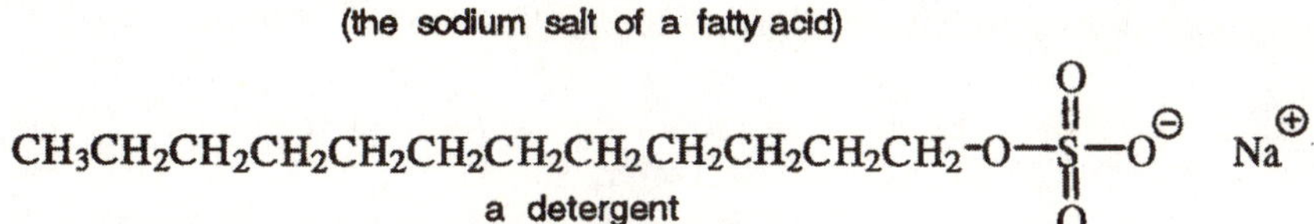

a detergent
(a sodium alkyl sulfate)

NONSAPONIFIABLE LIPIDS

Probably the most important lipid in this class, and the one you will work with in this experiment, is cholesterol. Cholesterol is a steroid. **Steroids** are compounds based on the structure of four rings shown on the next page.

Cholesterol is synthesized by animals and used as the starting compound to make other steroids such as hormones, several vitamins, and bile salts. Cholesterol, both produced by your body and from your diet, circulates in the blood in several forms in combination with protein. Heart disease appears to be linked with high blood levels of low-density lipoproteins (LDL or "bad" cholesterol) and low levels of high-density lipoproteins (HDL or "good" cholesterol). Food fats (triglycerides) raise blood cholesterol much more than does food cholesterol. Cholesterol is found in bile liquid, and is the principal component of gallstones.

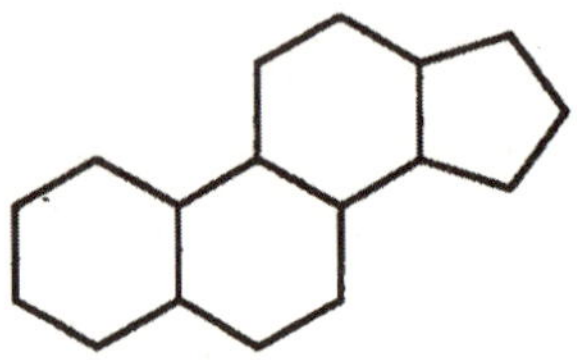

steroid nucleus

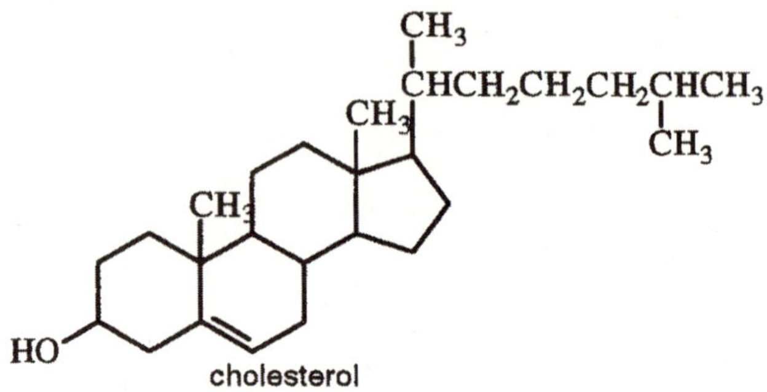

cholesterol

> **CAUTION: Most organic compounds are flammable. Ether, used in Part II, is EXTREMELY flammable. Keep away from all flames.**
>
> **Bromine vapors (Part IV) are very irritating to the eyes and lungs.**
>
> **Sodium hydroxide (Part V) is very corrosive (caustic) to animal and vegetable tissue, especially the hot, concentrated solution used in this experiment.**

Procedure

I. SOLUBILITY OF LIPIDS

1. Place 4 dry test tubes in a rack, number them, and add 1 mL of the following solvents. **CAUTION: No flames should be present.**
   - No. 1: Deionized water
   - No. 2: Ethanol
   - No. 3: Heptane
   - No. 4: Dichloromethane

2. Add about 10 drops of cottonseed oil to each tube, stopper, and shake vigorously. Record the solubility (S or I) on the Report Sheet.

3. Repeat the above procedure, first with commercial liquid shortening (vegetable oil) and then with lard instead of the cottonseed oil.

4. Disposal: Discard No. 2 (ethanol) and No. 3 (heptane) in the "Waste Organic Solvent" container. Discard No. 4 (dichloromethane) in the "Waste Halogenated Solvent" container.

II. KREIS TEST FOR OXIDATIVE RANCIDITY

This part may be performed as an instructor's demonstration.

1. **CAUTION: No flames should be present! Ether is EXTREMELY flammable! Perform under a fume hood!** Place about 3 mL of an "old" commercial shortening in one test tube, and an equal amount of fresh shortening in a second test tube.

2. Add 1 mL of concentrated hydrochloric acid (**CAUTION**); then stopper and shake each tube. Wash your hands.

3. Add 3 mL of phloroglucinol-ether solution to each test tube. Stopper and gently shake the contents once more and place the test tubes in a test tube rack for about 10 minutes. Ether evaporates rapidly (i.e., it is volatile); slight pressure may build up in the stoppered test tube. Wash your hands.

4. Remove each tube from the rack and add about 3 mL of ether to each. Stopper and gently shake each test tube. Place the tubes in a rack, and compare their appearances.

5. Disposal: Discard the solutions in the "Waste Organic Solvent" container.

III. IDENTIFICATION TEST FOR LIPIDS—ACROLEIN TEST

This part will be performed as an instructor's demonstration.

> **CAUTION: Do not begin this part until those working near you have completed Parts I and II. Ether is EXTREMELY flammable and MUST NOT be used near flames.**
>
> **Do this part under a fume hood.**

1. Place about 2 or 3 drops of cottonseed oil in a test tube, add a few crystals of potassium hydrogensulfate, $KHSO_4$.

2. Heat gently until a distinguishing odor is noticeable. **CAUTION: At this point stop heating the test tube.**

3. Repeat Steps 1 and 2, first substituting glycerol, then lard, and finally palmitic acid (a saturated fatty acid) for the cottonseed oil.

IV. UNSATURATION OF COMMERCIAL SHORTENING

CAUTION: Do this part in a fume hood. Wear disposable laboratory gloves.

A sample of lard, commercial solid shortening, and liquid commercial shortening will be assigned to you by your instructor. Use a different medicine dropper for measuring each sample. Liquefy solid shortenings just before they are to be used by placing them in a hot water bath on a hotplate and heat the test tube until the sample just completely melts. **CAUTION: Do not overheat.** Use the following procedure for each sample.

1. Fill a 25-mL buret with a 5% solution of bromine in dichloromethane. (**CAUTION!**)

2. Place 5 drops of the shortening to be tested in a 50-mL Erlenmeyer flask, and add 2 mL of dichloromethane. Place the flask on a white sheet of paper.

3. Add the bromine solution to the flask in 0.2-mL quantities, swirling the flask after each addition.

4. Continue these additions until a yellow color (due to unreacted bromine) persists for 30 seconds. Record the volume of bromine solution added to the sample.

5. Disposal: Discard the sample in the "Waste Halogenated Solvent" container. Discard the solution in the buret as directed by your instructor.

V. SAPONIFICATION OF A LIPID—PREPARATION OF SOAP (GROUP EXPERIMENT)

This procedure should be done only with constant stirring and at a controlled temperature. Therefore, the hotplate method using a magnetic stir bar is needed. Due to the use of hot alkali solution, this should also be done in a fume hood.

1. Place 10 g of lard in a 250-mL beaker, and add 40 mL of 50% sodium hydroxide solution. (**CAUTION, very caustic! Wear disposable laboratory gloves and wash your hands if any sodium hydroxide contacts your skin.**)

2. Place a magnetic stir bar in the mixture and put the beaker on a hotplate. Put an evaporating dish filled with ice on top of the beaker. Gently heat the mixture with stirring for about one half hour. Replace ice in the evaporating dish as needed.

3. Pour 50 mL of a saturated sodium chloride solution into a 400-mL beaker. Pour the hot solution from Step 2 into the salt solution. (**CAUTION!**) A layer of soap should form at the surface. The sodium chloride reduces the solubility of soap in water. Wash your hands.

4. Do not touch the strongly basic solution with bare skin. Wear gloves if necessary. Remove the soap, wash it with water, and save it for use in Part VI. Wash your hands.

VI. CLEANSING PROPERTIES OF SOAP AND DETERGENT

1. Prepare about 60 mL of soap solution from the small lumps of soap prepared in Part V, or pieces obtained from your instructor. Place 5 mL of soap solution into each of 5 numbered test tubes, and add the following reagents:
   No. 1: 5 mL of deionized water
   No. 2: 5 mL of tapwater
   No. 3: 5 mL of 0.1% CaC_{12} solution
   No. 4: 5 mL of 0.1% MgC_{12} solution
   No. 5: 5 mL of 0.1% NaCl solution

2. Stopper each of the test tubes and shake for about 30 seconds. Compare the foam formed in each case. Save test tube No. 3 ($CaCl_2$) for use in Step 6.

3. Repeat Steps 1 and 2 with a detergent solution instead of the soap solution.

4. Pour 5 mL of 0.1% $CaCl_2$ solution into a test tube, add about 3 g of ion-exchange resin, and stopper the test tube. Shake the test tube for about 3 or 4 minutes.

5. Filter the solution containing the resin, and collect the filtrate in a test tube.

6. Place 5 mL of soap solution (from Step 1) in the filtrate, stopper the test tube, and shake for about 30 seconds. Compare the foam formed in this case to the foam formed with the untreated $CaCl_2$ solution (Step 1—test tube No. 3).

7. Disposal: Discard the ion-exchange resin in the "Waste Ion-Exchange Resin" container.

VII. CHOLESTEROL

Examine samples of cholesterol or gallstones.

Name ______________________ Section ____________ Date ______________

Prelab Questions for Experiment 10

_______ 1. An unsaturated lipid
a. is more difficult to digest than a saturated lipid.
b. has a lower melting point than a saturated lipid.
c. spoils less easily than a saturated lipid.
d. is more soluble in polar solvents than a saturated lipid.

_______ 2. The acrolein test for a lipid is
a. based entirely on the fact that lipids can be saponified.
b. based upon the fact that certain lipids, upon heating, form glycerol, which is easily dehydrated.
c. applicable to nonsaponifiable lipids as well as saponifiable lipids.
d. specific for cholesterol.

_______ 3. If a commercial lipid produced a positive Kreis test, you should
a. return it to the store where you purchased it and demand a refund.
b. use it regularly for cooking.
c. use it as a salad dressing.
d. classify it as a nonsaponifiable lipid.

_______ 4. Cholesterol
a. is a saponifiable lipid.
b. is produced in the body as a result of the hydrolysis of a saponifiable lipid.
c. is produced in the body and used to make hormones.
d. is the principal component of kidney stones.

_______ 5. If a lipid is heated with $KHSO_4$ and acrolein is not produced, the lipid is probably
a. unsaturated.
b. saturated.
c. saponifiable.
d. nonsaponifiable.

6. Why is some water "hard"?

____

Name ______________________ Section ____________ Date ____________

Report for Experiment 10

I. SOLUBILITY OF LIPIDS

1. Data:

Lipid	*Solubility in Water (I or S)*	*Solubility in Ethanol (I or S)*	*Solubility in Heptane (I or S)*	*Solubility in Dichloromethane (I or S)*
Cottonseed oil				
Vegetable oil				
Lard				

2. Do the solubility data indicate lipids are polar or nonpolar? Explain.

____

____

II. KREIS TEST FOR OXIDATIVE RANCIDITY

1. Color observations and conclusions:

   a. Old commercial shortening ______________________

   b. Fresh commercial shortening ______________________

III. IDENTIFICATION TEST FOR LIPIDS—ACROLEIN TEST

1. Observations of odor:

   a. Cottonseed oil ______________________

   b. Glycerol ______________________

   c. Lard ______________________

   d. Palmitic acid ______________________

2. Explain any differences.

____

____

3. Write a chemical equation to illustrate the reaction of potassium hydrogensulfate with glycerol.

4. Use a word equation (or equations) to explain how the odor is produced from lard and cottonseed oil.

___

___

IV. UNSATURATION OF COMMERCIAL SHORTENING

1. Data:

Shortening	*Volume of Br2 solution*
Lard	
Commerical Salad Shortening	
Commercial Liquid Shortening	

2. a. Which of the shortenings tested is the most unsaturated?

   ___

   b. Does this agree with the manufacturer's claim?

   ___

V. SAPONIFICATION OF A LIPID—PREPARATION OF SOAP

1. Observations of the product of the saponification reaction.

___

___

2. Why was a salt solution used in the preparation of soap?

___

___

3. Write a word equation for the preparation of soap.

___

Name ________________ Section ________ Date ________

VI. CLEANSING PROPERTIES OF SOAP AND DETERGENT

1. Data: for Steps 1 through 3:

Solution added	*Amount of Foam Produced with Soap*	*Amount of Foam Produced with Detergent*
Deionized water		
Tapwater		
0.1% $CaCl_2$		
0.1% $MgCl_2$		
0.1% NaCl		

2. a. Which of the solutions tested produced the most foam with soap? ________

   b. With detergent? ________

3. a. Which of the solutions tested produced the least foam with soap? ________

   b. With detergent? ________

4. Which ions are usually present in hard water? ________

5. Data: Steps 4 through 6:

Solution added	*Amount of Foam Produced with Soap*
0.1% $CaCl_2$ (data from Step 1)	
0.1% $CaCl_2$ after ion-exchange treatment (Step 6)	

6. a. Did the ion-exchange treatment of the calcium chloride solution affect the amount of foam produced by the soap?

   ____

   b. Explain your result.

   ____

VII. CHOLESTEROL

1. Observations of cholesterol.

2. Observations of gallstones.

Name ________________________ Section ____________ Date ______________

Related Questions for Experiment 10

1. How does the degree of unsaturation relate to the digestibility of a shortening?

____

____

2. a. Which of the solvents in this experiment is an effective grease remover?

____

b. Explain why it is so effective.

____

____

3. The odor of acrolein is frequently strong in restaurants that feature charcoal-broiled food. Explain the presence of this odor.

____

____

4. How does a home water softener produce soft water?

____

____

5. During World War II, homemakers were urged to turn in bacon grease for use in the war effort. Explain how the grease could be converted into useful products.

____

____

6. The cleanser Zest® advertises that Zest does not leave a soap film. In what way is the chemical structure of Zest different from soap?

____

____

7. Classify the following fatty acids as saturated, monounsaturated, or polyunsaturated (S, M, or P):

_______ a. 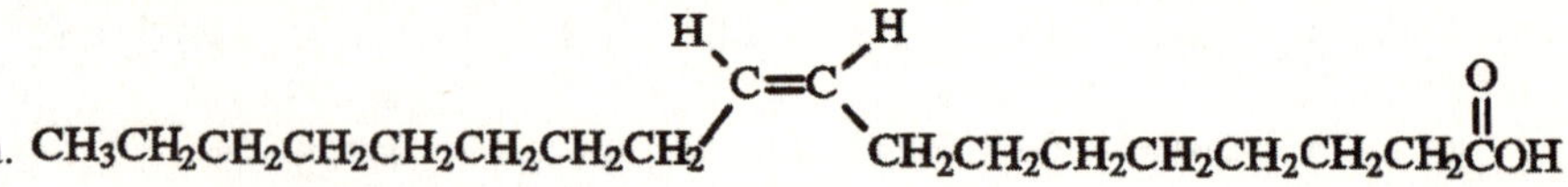

_______ b. $CH_3(CH_2)_{14}\overset{O}{\overset{\|}{C}}OH$

_______ c. 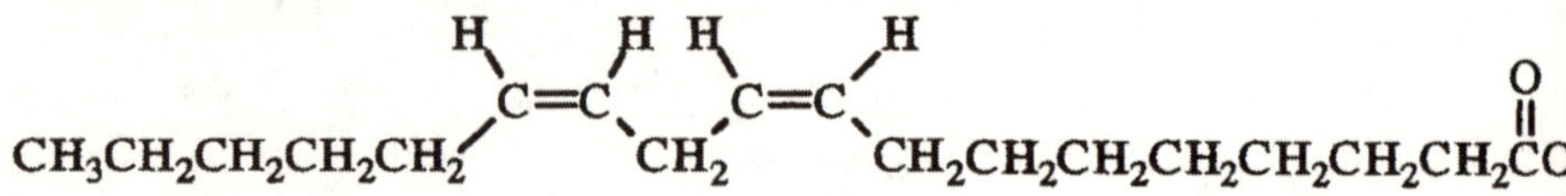

8. Classify the fatty acid in Question 7 as a *cis-* or *trans-* isomer.

____

9. Complete the following equation representing the hydrolysis of a triglyceride.

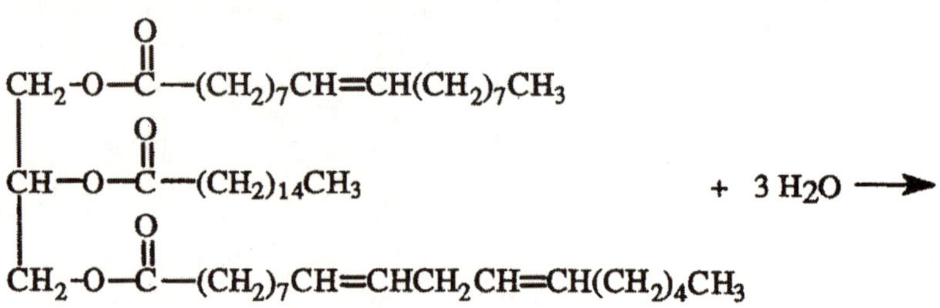

10. Complete the following equation representing the complete hydrogenation of a triglyceride.

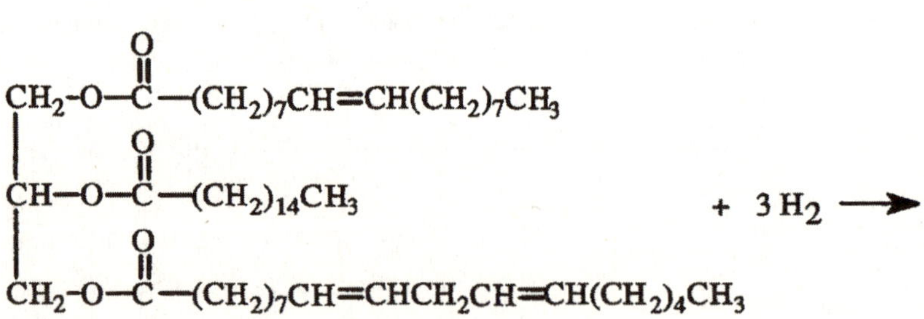

Experiment 11

Electrolytes, Acids, and Bases

References: *General, Organic, and Biological Chemistry: An Integrated Approach, Chapter 9.*

Purpose: You will study the behavior of acids, bases, and electrolytes.

Introduction

ELECTROLYTES AND NONELECTROLYTES

When a solute dissolves in a solvent, the solute particles may be charged ionic particles or uncharged molecular particles. Solutions of uncharged molecular particles do not conduct an electric current; substances that form uncharged molecular particles in aqueous solution are termed **nonelectrolytes**. Glucose, the sugar found in blood, exists in water as uncharged molecules and does not conduct an electric current.

$$\text{glucose}_{(s)} \xrightarrow{H_2O} \text{glucose}_{(aq)}$$

Aqueous solutions of **electrolytes** conduct an electric current because the solutions contain charged ionic particles. When dissolved in water, sodium chloride dissociates into sodium ions and chloride ions. The charged ions conduct an electric current; sodium chloride is an electrolyte.

$$NaCl_{(s)} \xrightarrow{H_2O} Na^{+}_{(aq)} + Cl^{1-}_{(aq)}$$

The electrolyte balance in body fluids plays a critical role in governing many processes in the human body ranging from controlling transport of ions across membranes to regulating the pH of blood. Electrolyte imbalances can pose serious health risks. Electrolyte ions are found in all body tissues and fluids and are routinely measured in blood serum and urine to evaluate acid-base and electrolyte balance. The most commonly measured serum electrolytes are sodium ion (Na^+, potassium ion (K^+), bicarbonate ion (HCO_3^-), and chloride ion (Cl^-). These ions are the most variable and frequently fluctuate in various disease states. Other principal electrolytes found in body fluids include calcium ion (Ca^{2+}), magnesium ion (Mg^{2+}), hydrogenphosphate ion (HPO_4^{2-}), sulfate ion (SO_4^{2-}), organic acid anions, and protein anions. **Anions** are negatively charged ions; positive ions are **cations**.

Some electrolyte solutions are good conductors of electricity—these are called strong electrolytes. Weak electrolytes are poor conductors of electric current. Both strong and weak electrolyte solutions contain ions; **strong electrolytes** are almost completely ionized in water and **weak electrolytes** are only slightly ionized. Nitric acid (HNO_3) is a strong electrolyte. In an equilibrium mixture of water and nitric acid, the equilibrium lies far to the right. More than 99 percent of the nitric acid ionizes to nitrate ions (NO_3^-); less than 1 percent of the nitric acid remains as uncharged molecules (HNO_3). Nitric acid is a strong electrolyte because it is almost totally ionized in water.

$$\underset{< 1\%}{HNO_{3\,(l)}} + H_2O \rightleftharpoons H_3O^+{}_{(aq)} + \underset{> 99\%}{NO^-_{3\,(aq)}}$$

Hydrofluoric acid (HF) is a weak electrolyte. The equilibrium lies far to the left in an equilibrium mixture of hydrofluoric acid and water. Only about 8 percent of the hydrofluoric acid ionizes to fluoride ion (F^-) and 92 percent is present as molecular HF. Because it is only slightly ionized, hydrofluoric acid solutions only weakly conduct electricity.

$$\underset{92\%}{HF_{(l)}} + H_2O \rightleftharpoons H_3O^+{}_{(aq)} + \underset{8\%}{F^-{}_{(aq)}}$$

The conductivity apparatus, illustrated in Figure 11-1, tests the electrical conductivity of solutions and is used to classify substances as strong electrolytes, weak electrolytes, or nonelectrolytes. Your instructor will perform the conductivity tests as a demonstration.

The conductivity tester is first plugged into an electrical outlet; the solution to be tested is poured into a clean beaker, and the electrodes are placed into the solution. A strong electrolyte will light a 60-watt bulb, a weak electrolyte will light a 15-watt bulb, and a nonelectrolyte will light neither bulb. Any degree of illumination will be assumed to be a positive test.

Several precautions must be observed in using the conductivity tester.

1. Electrodes must be washed with deionized water before testing a new solution. **Unplug the apparatus before washing.**

2. **To avoid severe electrical shock, do not touch the electrodes.**

3. When testing the solution, be sure the electrodes are immersed well beneath the surface of the solution.

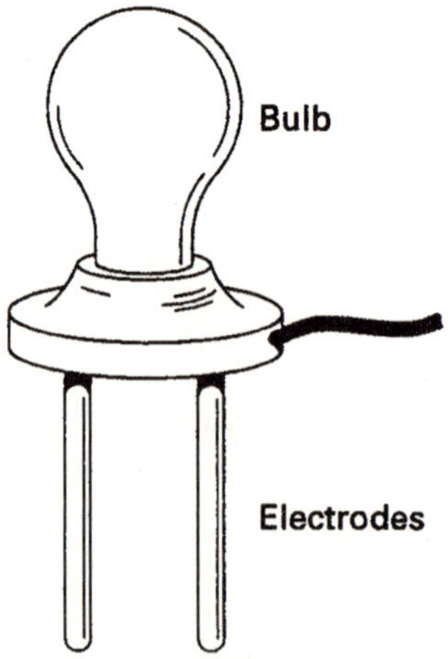

Figure 11-1 A typical conductivity tester.

ACIDS AND BASES

Acids and bases can be defined in several different ways. Generally, the most useful is the Bronsted-Lowry definition which states that an **acid** is a substance capable of donating a proton. A substance capable of accepting a proton is a **base**.

When added to water, all acids can donate a proton (H^+) to water; the products of the reaction will be a hydronium ion (H_3O^+) and a negative ion. Because ions are formed, all acids are electrolytes. **Strong acids** are almost completely ionized in water and are therefore strong electrolytes. Strong acids easily donate protons to acceptors. Nitric acid, mentioned earlier, is a strong acid. Only a small number of common acids are strong acids: hydrochloric acid (HCl), sulfuric acid (H_2SO_4), and nitric acid (HNO_3).

$$\underset{<1\%}{HNO_{3\,(l)}} + H_2O \rightleftharpoons \underset{>99\%}{H_3O^+_{(aq)} + NO^-_{3\,(aq)}}$$

Weak acids are only slightly ionized in water and are weak electrolytes. Weak acids donate protons, but with limited success. Hydrofluoric acid is a weak acid. Most common acids are weak acids.

$$\underset{92\%}{HF} + H_2O \rightleftharpoons \underset{8\%}{H_3O^+_{(aq)} + F^-_{(aq)}}$$

Bases, too, are either strong or weak. **Strong bases** are nearly totally ionized in water and hence are strong electrolytes; **weak bases** are weak electrolytes because they only partially ionize in water. Strong bases are better proton acceptors than weak bases. Sodium hydroxide (NaOH) and potassium hydroxide (KOH) are common strong bases; both ionize in water to release hydroxide ion (OH^-). The hydroxide ion is an excellent proton (H^+) acceptor.

$$OH^- + H^+ \longrightarrow H_2O$$

NEUTRALIZING REACTIONS

In Experiment 12, you will investigate one of the common reactions of acids—the reaction with a reactive metal to produce hydrogen gas. Another property of acids is that they react with and neutralize bases, producing a neutral solution containing a salt and water. Table salt (NaCl) is a salt but many other substances, such as KCl, are also salts. A **salt** is an ionic compound and is one of the products of an **acid-base neutralization reaction**. Because salts are ionic, they are electrolytes. All electrolytes are either an acid, a base, or a salt.

$$\text{acid} + \text{base} \longrightarrow \text{salt} + \text{water}$$

A solution of potassium hydroxide is basic. As hydrochloric acid is added to the solution, the acid consumes the base, converting it to a salt (KCl) and water. As the base is destroyed, the solution becomes neutral. If additional acid is added, the solution becomes acidic.

$$HCl + KOH \longrightarrow KCl + H_2O$$

Acids are neutralized by bases in the same manner. A solution of hydrochloric acid is, of course, acidic. If potassium hydroxide is added to the solution, the acid is consumed and the solution turns neutral. Additional potassium hydroxide will make the solution basic.

THE pH SCALE

The pH scale is a convenient way of indicating the acidity or basicity of aqueous solutions. Pure water and all neutral solutions contain equal concentrations of hydronium ions and hydroxide ions, $[H_3O^+] = [OH^-]$, and the pH is 7. An acidic solution is one in which the concentration of hydronium ions is greater than the concentration of hydroxide ions, $[H_3O^+] > [OH^-]$ and the pH is less than 7. When $[OH^-] > [H_3O^+]$, the solution is basic (alkaline) and the pH is greater than 7.

Buffers are solutions that resist a change in pH. They protect against drastic changes in pH when acids or bases are added to the solution. The body's blood and extracellular fluids contain buffers. The buffering capacity of a system may be overloaded if huge amounts of acid or base are added; in the blood, this can lead to acidosis or alkalosis.

The buffer used in this experiment is a mixture of potassium dihydrogenphosphate (KH_2PO_4) and potassium monohydrogenphosphate (K_2HPO_4). A solution of this mixture contains the following equilibrium:

$$H_2PO_4^- + H_2O \rightleftharpoons HPO_4^{2-} + H_3O^+$$

If acid is added to the buffer solution, the stress of added hydronium ion causes the equilibrium to shift to the left, relieving the stress. Most of the additional hydronium ion is changed into water and the pH stays relatively constant.

Bases decrease the concentration of hydronium ion in a solution. If base is added to a buffer solution, the stress of reduced hydronium ion concentration causes the equilibrium to shift to the right. Water is converted to hydronium ion and the pH does not significantly change.

A. The pH Meter

The pH meter provides the most accurate means of measuring the pH of a solution. pH meters are calibrated using standard buffer solutions; then the electrodes are dipped into the solution to be tested and a pH reading is made from the meter. Many different types of pH meters are available; your instructor will describe the use of your pH meter.

B. Indicators

Another simpler, but less accurate, method of measuring pH uses chemical dyes called indicators. **Indicators** change color at certain pH levels.

Bromthymol blue is a commonly used indicator that is yellow in acidic solution and blue in basic solution. In neutral solutions, this indicator is green. This experiment will use bromthymol blue as the indicator in a neutralization reaction.

Litmus is a naturally occurring dye that is either red or blue. Strips of **litmus paper** are commonly used to indicate whether a solution is acidic (pH < 7), basic (pH > 7), or neutral (pH = 7). Blue litmus paper turns pink when placed in an acidic solution, and red litmus paper turns blue in basic solution (remember this as *B=B*, if red paper turns **b**lue, the solution is **b**asic). If neither paper changes color, the solution is neutral.

Universal indicator, a mixture of dyes, is not used in this experiment but will be used in subsequent experiments. Several different mixtures of universal indicator are manufactured; the most common one displays the following colors when added to a solution.

pH	*Color*	*pH*	*Color*
4	Rose	8	Blue-green
5	Gold	9	Blue
6	Yellow	10	Violet
7	Green		

pH paper contains a mixture of indicators which turn a range of colors when wetted with a drop of solution. Each type of pH paper must be used in conjunction with the supplied table of colors to determine the pH of the tested solution. The most frequently used pH paper covers a wide pH range (1–12) but is accurate within only one pH unit. Other types of pH paper measure narrower pH ranges but will supply a greater degree of accuracy.

Procedure

I. ELECTROLYTES (Instructor's Demonstration)

1. Place 50 mL of each of the following solutions into numbered 100-mL beakers:
   No. 1: 0.9% salt (NaCl) solution
   No. 2: 1% sugar (sucrose, $C_{12}H_{22}O_{11}$) solution
   No. 3: Vinegar (5% acetic acid, $HC_2H_3O_2$)
   No. 4: 10% ethanol (C_2H_6O)
   No. 5: Deionized water
   No. 6: Tapwater
   No. 7: Milk of magnesia, a suspension of $Mg(OH)_2$
   No. 8: 0.1 M hydrochloric acid (stomach acid, HCl)
   No. 9: 0.1 M sodium hydroxide (lye, NaOH)
   No. 10: 0.1 M sodium bicarbonate (baking soda, $NaHCO_3$)

2. Test each of the solutions with the conductivity tester as described in the Introduction. Classify each substance as a strong electrolyte (S), weak electrolyte (W), or nonelectrolyte (N).

3. Place 25 mL of glacial (100%) acetic acid into a 100-mL beaker and test its conductivity. While monitoring the conductivity, gradually add 50 mL of deionized water to the beaker. Note the results. (**CAUTION: Glacial acetic acid is corrosive and has a very strong odor. Adding water to concentrated acid is generally an unsafe procedure—this particular addition may be done with caution.**)

II. MEASUREMENT OF pH

A. Acidity and Basicity of Substances

1. Label ten 10 × 75-mm test tubes (smallest size) from 1 to 10.

2. Obtain about 1 mL of each of the solutions in Step 1 of Part I.

3. Test the acidity and basicity of each solution by dipping a stirring rod into the solution, and then placing the tip of the rod on red and blue litmus paper. Wash and dry the rod after each test.

B. pH of Foods

1. Pour about 1 mL of each of the following solutions into numbered 10 × 75-mm test tubes:
   No. 1: Orange juice
   No. 2: Tomato juice
   No. 3: Lemon juice
   No. 4: Tea
   No. 5: A carbonated beverage

2. Use a small piece, only 1 cm (1/2 inch) long, of wide-range indicator paper (pH range 1–12) to measure the pH of the solutions. With a stirring rod, place one drop of the solution on the pH paper. Use the color chart to estimate the pH. Be sure to rinse the stirring rod with deionized water and dry after each measurement.

C. pH of a Carbonic Acid Solution

1. Use either distilled water or boiled deionized water in this procedure. Ordinary deionized water contains dissolved carbon dioxide; heating it to boiling removes the dissolved carbon dioxide. The deionized water must be boiled in advance and allowed to cool to room temperature.

2. Pour 50 mL of distilled water or boiled deionized water into a 100-mL beaker. If possible, use a pH meter to record the pH. Because of the absence of ions, pH meter measurements of distilled water are usually unstable. If no pH meter is available, use narrow-range pH paper.

3. Blow through a straw into the water for approximately five minutes. The carbon dioxide of your exhaled breath will react with the water and form carbonic acid. Record the final pH of the water.

III. ACID-BASE NEUTRALIZATION

1. Pour about 30 mL of 0.1 M sodium hydroxide (NaOH) solution into a labeled 100-mL beaker.

2. Pour about 30 mL of 0.1 M hydrochloric acid (HCl) solution into a labeled 100-mL beaker.

3. Transfer 1 mL of the 0.1 M NaOH to a 16 × 150-mm test tube.

4. Add one drop of bromthymol blue solution to the test tube. Record the color of the solution.

5. While stirring the solution, add 0.1 M HCl from an eye dropper one drop at a time. Count and record the drops added until the solution just turns green or yellow.

6. Add one or two additional drops of HCl and record the color.

7. Using a clean eye dropper, add 0.1 M NaOH to the test tube, one drop at a time, until the solution just turns green or blue.

8. Add one or two additional drops of NaOH and note the color.

9. Repeat Steps 5 through 8. The solution is neutral when the solution is green.

10. Save the NaOH and HCl solutions for use in Part IV.

IV. BUFFERS

A. The Buffering Ability of Water

1. Review Fundamental Laboratory Operation, Part 3c, page 12, for the preparation and use of the buret.

2. Clean two 25-mL burets with a buret brush, rinse well with tap water, allowing the water to drain through the tip. Rinse with a few milliliters of deionized water.

3. Place both burets in a buret clamp. Label one buret "NaOH" and the other "HCl."

4. Using the 0.1 M NaOH solution from Part III, pour a few milliliters of NaOH solution into the first buret. Discard the solution through the tip since the solution is diluted with rinse water.

5. Pour about 20 mL of 0.1 M NaOH into the buret. Allow a small amount of NaOH solution to drain through the tip until there is a solid column of solution extending down to the tip, completely free of air bubbles.

6. Adjust the level of NaOH solution to the 15-mL mark on the buret. Remember to read the bottom of the meniscus.

7. Pour 25 mL of distilled water or boiled deionized water into a 150-mL beaker. Use a pH meter to record the pH.

8. Use the buret to add 0.5 mL of NaOH to the beaker of deionized water. Stir the mixture. Record the pH.

9. Add the NaOH in 0.5 mL quantities until a volume of 5 mL has been added. Stir after each addition and record the pH.

10. Discard the solution in the beaker, wash and dry it thoroughly. Thoroughly rinse the electrodes of the pH meter.

11. Repeat steps 4 through 10 using 0.1 M HCl in place of NaOH.

12. Save the 0.1 M solutions of NaOH and HCl in the burets for Part B.

B. The Buffering Ability of a Phosphate Buffer

1. Pour 25 mL of a KH_2PO_4/K_2HPO_4 buffer solution into a 150-mL beaker. Record the pH using a pH meter.

2. Use the buret to add 0.1 M NaOH in 0.5 mL quantities until a volume of 5 mL has been added. Stir after each addition and record the pH.

3. Discard the solution in the beaker, wash and dry it thoroughly. Thoroughly rinse the electrodes of the pH meter.

4. Pour 25 mL of a KH_2PO_4/K_2HPO_4 buffer solution into a 150-mL beaker. Record the pH using a pH meter.

5. Use the buret to add 0.1 M HCl in 0.5 mL quantities until a volume of 5 mL has been added. Stir after each addition and record the pH.

6. Discard the solutions from the burets. Rinse the burets thoroughly with tap water, allowing the water to drain through the tip. Rinse with a few mL of deionized water.

7. Thoroughly rinse the electrodes of the pH meter.

C. The Buffering Ability of Alka-Seltzer®

1. Pour exactly 200 mL of deionized water into a 250-mL beaker. Add an Alka-Seltzer® tablet and stir the mixture. Use a pH meter to record the pH.

2. Measure 100 mL (half of the above solution) into a second 250-mL beaker.

3. Fill a 25-mL buret to the zero mark with 0.1 M NaOH and fill a second 25-mL buret to the zero mark with 0.1 M HCl.

4. Add the NaOH to one of the Alka-Seltzer solutions in 5-mL quantities until a volume of 25 mL has been added. Record the pH after the addition of each increment. Be sure to stir the solution before taking a measurement.

5. Discard the solution in the beaker, wash and dry it thoroughly, and repeat Steps 1 through 3 with the second Alka-Seltzer solution using 0.1 M HCl in place of NaOH.

6. Turn off the pH meter. Thoroughly rinse the electrodes.

Name ______________________________ Section ____________ Date ______________

Prelab Questions for Experiment 11

________ 1. The electrical conductivity of a solution is tested with a conductivity tester. The solution fails to light a 60-watt bulb but does illuminate a 15-watt bulb. The solution is
a. a weak electrolyte.
b. a strong electrolyte.
c. a nonelectrolyte.

________ 2. Which of the following will provide the most accurate means of measuring pH?
a. bromthymol blue.
b. litmus paper.
c. pH paper.
d. a pH meter.

________ 3. When a drop of a solution is placed on red litmus paper, the paper turns blue. The solution is
a. acidic.
b. basic.
c. neutral.

________ 4. A solution becomes yellow after the addition of one drop of bromthymol blue. The solution is
a. acidic.
b. basic.
c. neutral.

________ 5. After adding 5 mL of acid to a solution having a pH of 9.6, the pH changed to 9.5. The solution is
a. acidic.
b. a buffer.
c. a nonelectrolyte.
d. a weak electrolyte.

________ 6. Which statement about a pH 5 solution is correct?
a. The solution is neutral.
b. The solution is a buffer.
c. The hydroxide ion concentration is greater than the hydronium ion concentration.
d. The hydronium ion concentration is greater than the hydroxide ion concentration.

Name ______________________ Section ____________ Date ______________

Report for Experiment 11

I. ELECTROLYTES (Instructor's Demonstration)

1. Data

	Solution	*Conductivity (S, W, or N)*	*Acid or Base (A, B, or N)*
1	0.9% NaCl		
2	1% sugar ($C_{12}H_{22}O_{11}$)		
3	Vinegar (5% $HC_2H_3O_2$)		
4	10% ethanol (C_2H_6O)		
5	Deionized water		
6	Tapwater		
7	Milk of magnesia, $Mg(OH)_2$		
8	0.1 *M* HCl		
9	0.1 *M* NaOH		
10	0.1 *M* $NaHCO_3$		

Glacial (100%) acetic acid	
Acetic acid diluted with H_2O	

2. Suggest a reason for the differing conductivities of tapwater and deionized water.

____

3. a. Complete the equation for the acid-base reaction between acetic acid and water.

$$HC_2H_3O_2 + H_2O \rightleftharpoons$$

b. Explain, with use of the equation above, the differing conductivities of 100% acetic acid ($HC_2H_3O_2$) and diluted acetic acid.

____

____

II. MEASUREMENT OF pH

A. Acidity and Basicity of Substances

Use the table in Part I to record the results of the litmus tests.

B. pH of Foods

1. Data:

	Solution	*pH*
1	Orange juice	
2	Tomato juice	
3	Lemon juice	
4	Tea	
5	Carbonated beverage	

2. Do the tested foods tend to be acidic or basic?

C. pH of a Carbonic Acid Solution

1. Data:

Initial pH of deionized water ______________

Final pH of deionized water
(after blowing through straw) ______________

2. Based on the equation $CO_2\ (g) + H_2O\ (l) \rightleftharpoons H_2CO_3\ (aq)$, explain the pH change that occurred when you blew through a straw into the deionized water.

____

____

III. ACID-BASE NEUTRALIZATION

1. Data:

Color of bromthymol blue indicator in NaOH solution (Step 4) ____________________

Number of drops of HCl to turn solution green (Step 5) ____________________

Color of solution with 1 or 2 additional drops of HCl (Step 6) ____________________

Name ______________________________ Section ______________ Date ________________

2. Is the color change reversible (can you change the color back and forth from yellow to blue)?

____

3. Write a complete, balanced equation for the neutralization of sodium hydroxide by hydrochloric acid.

____

IV. BUFFERS

A. The Buffering Ability of Water

1. Data:

mL of NaOH	*pH*		*mL of HCl*	*pH*
0.0			0.0	
0.5			0.5	
1.0			1.0	
1.5			1.5	
2.0			2.0	
2.5			2.5	
3.0			3.0	
3.5			3.5	
4.0			4.0	
4.5			4.5	
5.0			5.0	

2. A buffer is a substance that resists changes in pH when a small amount of acid or base is added. Is water an effective buffer?

____

B. The Buffering Ability of a Phosphate Buffer

1. Data:

mL of NaOH	*pH*		*mL of HCl*	*pH*
0.0			0.0	
0.5			0.5	
1.0			1.0	
1.5			1.5	
2.0			2.0	
2.5			2.5	
3.0			3.0	
3.5			3.5	
4.0			4.0	
4.5			4.5	
5.0			5.0	

2. Is the KH_2PO_4/K_2HPO_4 mixture a more effective buffer than water?

____

3. Was there evidence that the buffering capacity was used up by addition of large amounts of acid or base? Explain.

____

____

4. Label the conjugate acid-base pairs in the acid-base equilibrium of the buffer system:

$$H_2PO_4^- + H_2O \rightleftharpoons HPO_4^{2-} + H_3O^+$$

Name ______________________ Section ____________ Date ______________

C. The Buffering Activity of Alka-Seltzer®

1. Data:

mL of NaOH	*pH*		*mL of HCl*	*pH*
0.0			0.0	
5.0			5.0	
10.0			10.0	
15.0			15.0	
20.0			20.0	
25.0			25.0	

2. Does Alka-Seltzer® seem to be more effective in resisting changes in pH when acid is added or when base is added?

____

____

3. How does the buffering action of Alka-Seltzer® compare with that of KH_2PO_4/K_2HPO_4 buffer?

____

____

Name ______________________ Section ____________ Date ______________

Related Questions for Experiment 11

1. Acidosis is a condition in which the blood pH drops below normal. Acidosis can be caused by metabolic or respiratory conditions. Explain the role of carbon dioxide in producing respiratory acidosis. Hint: review this experiment and consult your text. (Chapter 9, Section 9.10, Health Link, in General, Organic, and Biological Chemistry: An Integrated Approach.)

2. a. Would a person who is hyperventilating suffer from alkalosis or acidosis?

   b. Explain.

   c. Explain why breathing into a paper bag is a treatment for hyperventilation.

3. An upset stomach can be caused by excess stomach acid. Why can baking soda ($NaHCO_3$) settle an upset stomach?

4. a. Which substance, H2P$_4^-$ or HPO_4^{2-}, would appear in the urine in alkalosis?

   b. In acidosis?

   c. Explain.

5. A student measured the pH of coffee and obtained the following results:
   caffeinated coffee pH 5.6
   decaffeinated coffee pH 5.0

   a. On the basis of this data, would you classify caffeine as an acid or base? __________

   b. Explain.

   ____

   ____

6. Bacteria in the mouth convert sugar into acids that are capable of dissolving tooth enamel. Repeated drinking of carbonated beverages, even those that do not contain sugar, can also contribute to loss of enamel. Explain why carbonated beverages can act in this way and cause dental caries (tooth decay).

   ____

   ____

7. Classify each of the following acids as strong or weak acids (S or W):

   ________ HNO_3

   ________ H_3PO_4

   ________ H_2CO_3

   ________ citric acid

   ________ sulfuric acid

   ________ ascorbic acid

8. Complete the following table to illustrate the difference between acid strength and acid concentration:

	Strong or Weak Acid? *(S or W)*	*Concentrated or Dilute Acid?* *(C or D)*
12 M HCl (hydrochloric acid)		
0.1 M HCl		
17 M $HC_2H_3O_2$ (glacial acetic acid)		
0.1 M $HC_2H_3O_2$		

Experiment 12

Chemical Reactions

Reference: *General, Organic, and Biological Chemistry: An Integrated Approach, Chapter 6.*

Purpose: You will perform several reactions to investigate factors that affect the rate of a chemical reaction. Chemical equilibrium and Le Châtelier's Principle will be studied. The method of continuous variation will be used to verify a balanced equation.

Introduction

REACTIONS RATES

The **rate** of a chemical reaction is the speed at which a reaction occurs. Some reactions, like the rusting of an iron nail, take years to complete. Other reactions, like the explosion of dynamite, are over in a fraction of a second.

In order to react, two molecules must physically collide with each other; the more frequent the collisions, the faster the reaction. The molecules must collide with enough energy (the **activation energy**) to overcome the repelling forces between the molecules. If molecules do not collide with energy at least as large as the activation energy, they will not react. The more collisions meeting or exceeding the activation energy, the faster the reaction.

You will investigate several factors that affect the rate of a chemical reaction: the nature of the reactants, the concentration of the reactants, the surface area of solid reactants, and the temperature. These factors determine the frequency and energy of reactant collisions.

In this experiment you will study the reaction rates of acids with metals under different conditions. You will vary the concentration of reactants by using acids of differing molarity. **Molarity** (M) is defined as the number of moles of solute per liter of solution. If two solutions have equal volumes, the one having a greater molarity will have a greater number of solute molecules.

You will investigate the reaction of hydrochloric acid (HCl) with the metals copper and zinc. The nature of the metals will determine *if* the metal reacts and how fast it reacts. *If* the metals react, they will produce a water-soluble metal chloride along with hydrogen gas. The unbalanced equations are:

$$Cu_{(s)} + HCl_{(aq)} \longrightarrow CuCl_2\,(aq)) + H_{2\,(g)}$$

$$Zn_{(s)} + HCl_{(aq)} \longrightarrow ZnCl_2\,(aq)) + H_{2\,(g)}$$

It is possible that the metal will not react:

$$Cu_{(s)} + HCl_{(aq)} \longrightarrow \text{No Reaction}$$

$$Zn_{(s)} + HCl_{(aq)} \longrightarrow \text{No Reaction}$$

In the above equations, (*s*) represents a solid, (*g*) a gas, (*l*) a liquid, and (*aq*) an aqueous solution.

When a metal reacts with an acid, vigorous bubbling, due to the liberation of hydrogen gas, takes place at the surface of the metal. The presence of hydrogen gas bubbles is evidence that a reaction is occurring. The presence of hydrogen can be verified by placing a burning wooden splint above the surface of reactants. A whistling or popping sound indicates the presence of hydrogen. The lighted splint furnishes the activation energy needed to initiate the reaction between hydrogen and oxygen.

$$\text{hydrogen} + \text{oxygen} \longrightarrow \text{water}$$

The intensity of the sound is determined by the concentration of oxygen and hydrogen. Pure hydrogen will produce a muffled "whoof," while the right mixture of hydrogen and air gives a loud "pop." Thus, the intensity of the sound of the reaction between hydrogen and oxygen cannot be used to judge the rate of the reaction between the metal and hydrochloric acid. The rate of reaction between metal and hydrochloric acid can be qualitatively judged by observing how rapidly the metal produces bubbles of gas. A more quantitative determination can be made by timing the number of minutes before the metal disappears.

CAUTION: Acids are corrosive and can severely damage tissue. Wear safety goggles while handling acids. If acid is spilled, flood the area with water and notify the instructor immediately. If acid contacts with skin, remove any clothing covering the skin and flush with large amounts of water. If acid contacts the eyes, flush with water for 20 minutes. Wash your hands after handling acids.

CATALYSIS

A **catalyst** is a substance that increases the rate of a chemical reaction, not by increasing the number or energy of molecular collisions, but by lowering the activation energy required for the reaction. The catalyst interacts with the reactants to produce an alternative path or route for the reaction—a route with a lower activation energy. Reactant molecules are more likely to collide with sufficient energy and the reaction rate is greater in the presence of a catalyst.

The reaction under study in this experiment is the decomposition (breaking apart) of hydrogen peroxide, H_2O_2. Hydrogen peroxide slowly decomposes as it sits in its bottle, gradually turning to water and oxygen gas. The unbalanced equation:

$$H_2O_2\,(aq) \longrightarrow H_2O\,(l) + O_2(g)$$

This reaction rate is accelerated by a catalyst. The presence of oxygen, which indicates that decomposition has taken place, can be easily verified. A glowing wooden splint is slowly burning in the oxygen of air. If the glowing splint is placed in a higher concentration of oxygen, the rate of combustion will increase and the wood splint will burst into flames.

Hydrogen peroxide is an **oxidizing agent**—it removes electrons from other substances. It destroys organic compounds that come in contact with it. A 3% solution of hydrogen peroxide is used as an antiseptic for small cuts of the skin—it oxidizes germs to death. When placed on an open wound, the hydrogen peroxide bubbles; the bubbles are oxygen gas. The exposed blood contains an enzyme (peroxidase) that catalyzes the decomposition of hydrogen peroxide. The oxygen bubbles come from the hydrogen peroxide, not from the catalyst. A catalyst is neither used up in a reaction nor irreversibly altered; a catalyst is a condition of the reaction and, if included in an equation, is written above the equation arrow. The unbalanced equation:

$$H_2O_{2\,(aq)} \xrightarrow{\text{peroxidase}} H_2O_{\,(l)} + O_{2\,(g)}$$

When a red spot is suspected of being blood, forensic chemists can use this reaction in a presumptive test for blood. The release of oxygen from hydrogen peroxide by the peroxidase activity causes an indicator to change color. This is not a definitive test for blood since other substances can also catalyze the decomposition of hydrogen peroxide.

Hydrogen peroxide is used in higher concentrations (6%) to bleach (oxidize) hair pigments in many hair-dyeing systems. Still higher concentrations (90%) are sometimes used as oxidizers in rocket propulsion.

CAUTION: Solutions of hydrogen peroxide, especially greater than 3%, should be handled with great care. If spilled, immediately flood the area with water and wash the liquid down the drain. Be careful not to spill any on your person, books, or clothing. Concentrations greater than 3% may cause white patches or blisters and chemical burns on the skin.

CHEMICAL EQUILIBRIUM AND LE CHATELIER'S PRINCIPLE

Some chemical reactions, like the reaction of metal with acid or the decomposition of hydrogen peroxide, "go to completion"—that is, every molecule of reactant is eventually converted to product. Other chemical reactions do not go to completion—the reactant is never totally changed into product. The later type of reaction is a chemical equilibrium. Chemical equilibriums do not go to completion because of a reverse reaction that competes with the forward reaction. In an **equilibrium**, the forward reaction occurs at the same rate as the reverse reaction.

Let's consider a hypothetical chemical reaction in which the reactants are represented by A and B, and the products by C and D. At the beginning of the reaction, only A and B are present, but as the reaction proceeds, C and D are formed. As the concentrations of A and B decrease, the concentrations of C and D increase. The forward reaction may be represented by the equation:

$$A + B \longrightarrow C + D$$

If C or D is not removed from the reaction site, they can react to form the original substances A and B:

$$C + D \longrightarrow A + B$$

As the concentrations of A and B decrease, the forward reaction slows down. As the concentrations of C and D increase the reverse reaction speeds up. Eventually the rate of both reactants becomes equal and an equilibrium is established:

$$A + B \rightleftharpoons C + D$$

In other words, after a period of time, a mixture of products and reactants exists. On the microscopic level, A and B are constantly changing to C and D in the forward reaction and C and D are constantly changing to A and B in the reverse reaction. However, the concentrations of A, B, C, and D remain constant because the forward and reverse reactions occur at the same rate. The composition of this equilibrium mixture will remain constant unless the concentrations of reactants or products are changed by the experimenter. Increasing the concentration of A or B causes the forward reaction to proceed faster, forming additional C and D, until the rate of the reverse reaction increases enough to again establish a state of equilibrium. In the new equilibrium the concentrations of C and D, represented by [C] and [D], are greater than in the original equilibrium.

The nineteenth century French chemist Henri Louis Le Châtelier described this increased concentration of A or B as a "stress." **Le Châtelier's Principle** states that if stress is applied to an equilibrium system, the system will respond by altering the equilibrium in such a way as to minimize the stress. Increasing the concentration of A or B is a stress that can be relieved by removing some of the added material and converting A and B to C and D. The equilibrium "shifts to the right" and a new equilibrium is established with higher concentrations of C and D.

Similarly, addition of C or D to a system at equilibrium will cause the equilibrium to "shift to the left" and the new equilibrium position will have higher concentrations of A and B than the original equilibrium. You can also cause equilibrium shifts by removing reactants or products (decreasing their concentrations). Thus, removal of A or B from the equilibrium system will cause a shift to the left. Removal of C or D will cause a shift to the right.

In this experiment you will examine two equilibrium systems:

$2\ K_2CrO_4\ (aq)$	+	$2\ HCl\ (aq)$	$\rightleftharpoons$	$K_2Cr_2O_7\ (aq)$	+	H_2O	+	$2\ KCl\ (aq)$
potassium chromate (yellow)		hydrochloric acid (colorless)		potassium dichromate (orange)		water (colorless)		potassium chloride (colorless)

$FeCl_3$	+	$3\ NH_4CNS\ (aq)$	$\rightleftharpoons$	$Fe(CNS)_3\ (aq)$	+	$3\ NH_4Cl\ (aq)$
iron(III) chloride (pale yellow)		ammonium thiocyanate (colorless)		iron(III) thiocyanate (deep red)		ammonium chloride (colorless)

The shifts in equilibrium can be easily followed in each case because the color of reactants is different than the color of products. In the chromate/dichromate equilibrium, when the equilibrium shifts to the right, the mixture becomes more orange-colored; when the equilibrium shifts to the left, the mixture becomes more yellow-colored. The concentration of a reactant will be increased by adding hydrochloric acid. The concentration of a reactant will be decreased by adding sodium hydroxide—sodium hydroxide is a base and it neutralizes, or removes, hydrochloric acid.

EXPERIMENTALLY VERIFYING A CHEMICAL EQUATION

Consider the unbalanced equation representing a reaction carried out in aqueous solution:

$$CaCl_{2\,(aq)} + AgNO_{3\,(aq)} \longrightarrow Ca(NO_3)_2 + AgCl\downarrow$$

This reaction is not an equilibrium—it goes to completion. One of the products, silver chloride (AgCl), is insoluble in water and, therefore, forms a precipitate ($\downarrow$). The following procedure, called **continuous variation**, illustrates a laboratory method for determining the number of moles of $AgNO_3$ that reacts with $CaCl_2$. The number of moles of each reactant is also, of course, the coefficient in the balanced equation. The experimental results should be the same as the results obtained by trial-and-error balancing.

Imagine that aqueous solutions of calcium chloride, $CaCl_2$, and silver nitrate, $AgNO_3$, are prepared so that each has the same concentration (molarity). One milliliter of $CaCl_2$ solution contains the same number of moles as one milliliter of $AgNO_3$ solution. Eight test tubes are filled with the solutions indicated in the following table (Figure 12-1). Each test tube contains a total of 9 mL of added solution, but the quantities of each solution are continuously varied in a systematic manner. The reaction produces a precipitate (AgCl) in each of the nine test tubes, and the solid precipitate settles to the bottom of each tube. The height of the precipitate is then noted. Why does test tube No. 3 have the most precipitate?

Test Tube No.	*mL of 0.1 M $CaCl_2$*	*mL of 0.1 M $AgNO_3$*	*Height of AgCl precipitate (mm)*
1	1	8	9
2	2	7	19
3	3	6	28
4	4	5	23
5	5	4	19
6	6	3	14
7	7	2	9
8	8	1	5

Figure 12-1 Continuous variation in data.

In test tube 3, all of the calcium chloride reacted with all of the silver nitrate—each reactant was completely used up. In test tubes 1 and 2, all of the calcium chloride reacted but not all the silver nitrate reacted. The reaction ended when the calcium chloride was gone and excess silver nitrate remained unreacted. In test tubes 4 through 8, all the silver nitrate reacted, but not all the calcium chloride reacted—the reaction stopped when the silver nitrate was used up.

Only test tube 3 had the proper amounts of both reactants. Calcium chloride and silver nitrate must react in a 3 to 6 molar ratio since that test tube contained 3 mL and 6 mL of the two reactants.

Reducing the 3 to 6 ratio gives the smallest whole number ratio of 1 mole of calcium chloride to 2 moles of silver nitrate. These are the coefficients of the reactants in the partially balanced equation:

$$1\ CaCl_2\ (aq) \ + \ 2\ AgNO_3\ (aq) \longrightarrow Ca(NO_3)_2\ (aq) \ + \ AgCl \downarrow$$

Of course, coefficients of 1 are not written in equations:

$$CaCl_2\ (aq) \ + \ 2\ AgNO_3\ (aq) \longrightarrow Ca(NO_3)_2\ (aq) \ + \ AgCl \downarrow$$

This method of continuous variation does not provide the coefficients of the products.

In this experiment, you will study the reaction of potassium chromate with barium chloride in the following unbalanced equation:

$$K_2CrO_4\ (aq) \ + \ BaCl_2\ (aq) \longrightarrow KCl\ (aq) \ + \ BaCrO_4 \downarrow$$

potassium chromate + barium chloride ⟶ potassium chloride + barium chromate

The method of continuous variation will be used to establish the molar ratio of the reactants, K_2CrO_4 and $BaCl_2$.

REACTION OF A PENNY WITH HYDROCHLORIC ACID (Optional)

U.S. pennies made in 1981 or earlier years were made from almost pure copper. Pennies made in 1983 and recent years have a zinc core which is coated with a thin layer of copper. Both types of pennies were made in 1982.

You will investigate one of the newer pennies with a zinc core. By filing away a little of the copper coating, you can expose the zinc core. Which metal will react with hydrochloric acid when the penny is placed in hydrochloric acid? The reaction is too slow to finish in one laboratory period. It is best if you can finish the procedure 24 hours later. The procedure may not work as well if you wait more than 24 hours.

The penny will be destroyed by this procedure. It is not illegal to destroy a penny or any other U.S. legal tender. It is illegal to make a penny look like a dime and attempt to spend it—that is counterfeiting.

Procedure

I. FACTORS THAT AFFECT THE RATE OF A CHEMICAL REACTION

A. Nature of the Reactants

CAUTION: Acids are corrosive and can severely damage tissue. Wear safety goggles while handling acids. If acid is spilled, flood the area with water and notify the instructor immediately. If acid contacts with skin, remove any clothing covering the skin and flush with large amounts of water. If acid contacts the eyes, flush with water for 20 minutes. Wash your hands after handling acids.

1. Place 5 mL of 6 M hydrochloric acid, HCl, into each of two test tubes. Use **CAUTION** when handling hydrochloric acid. Place the tubes in a test tube rack. Wash your hands after handling HCl.

2. Cut a strip of polished zinc and a strip of polished copper of equal size, about 1 cm × 1 cm.

3. Place the zinc in one of the test tubes and the copper in the other. Carefully observe each tube. Use a one-hole rubber stopper to loosely stopper each of the tubes for 30 seconds.

4. Remove the stoppers and test for the presence of hydrogen by placing a burning wood splint at the mouth of each of the two test tubes.

5. Clean-up: Fill the test tubes with tap water and then discard the liquid of the test tubes into the sink. Try to keep any metal pieces in the test tube. Again fill the test tube with water and discard the liquid. Transfer the metal piece to the "Waste Metal" container or as directed by your instructor. Use this same clean-up procedure in Parts B, C, and D.

B. Concentration of the Reactants

1. Pour 5 mL of 6 M hydrochloric acid into one test tube and 5 mL of 1 M hydrochloric acid into a second test tube. Place the tubes in a test tube rack.

2. Cut two pieces of zinc metal, each with a mass of about 0.2 gram, and insert a piece into each of the two test tubes. Note the time of day the zinc is added to the hydrochloric acid. Observe the relative rate of gas bubble formation in each tube.

3. Loosely stopper each test tube with a one-hole stopper for about 30 seconds and test for the presence of hydrogen with a burning splint.

4. Note the time of day when the reaction ceases (when no zinc remains).

C. Surface Area of a Solid Reactant

1. Weigh approximately 0.2 gram of zinc sheet metal. Place the metal into a clean, dry test tube.

2. Weigh an equal quantity of powdered zinc, and pour it into a second clean, dry test tube.

3. Add exactly 3 mL of 3 M hydrochloric acid to each of the two test tubes. Note the time of day the hydrochloric acid is added to the zinc. Observe each test tube.

4. Loosely stopper each test tube with a one-hole stopper for about 30 seconds and test for the presence of hydrogen with a burning splint.

5. Note the time of day when the reaction ceases (when no zinc remains).

D. Temperature of the Reactants

1. Prepare an ice-water mixture (bath) in a 250-mL beaker.

2. Pour 5 mL of 6 M hydrochloric acid into each of the test tubes. Place one of the test tubes in the ice-water bath and the other in a test tube rack. Wait 10 minutes for the test tube in the ice bath to cool.

3. Cut two pieces of zinc metal, each with a mass of about 0.2 gram.

4. Simultaneously drop the zinc into each of the two test tubes. Note the time of day the zinc is added to the hydrochloric acid. Observe each test tube. The test tube in the ice bath must remain in the ice bath.

5. Note the time of day when the reaction ceases (when no zinc remains).

II. CATALYSIS

CAUTION: Solutions of hydrogen peroxide, especially greater than 3%, should be handled with great care. If spilled, immediately flood the area with water and wash the liquid down the drain. Be careful not to spill any on your person, books, or clothing. Concentrations greater than 3% may cause white patches or blisters and chemical burns on the skin.

1. Pour 10 mL of 3% hydrogen peroxide (**CAUTION!**) into a test tube. Wash your hands after handling hydrogen peroxide.

2. Loosely stopper the test tube with a one-hole stopper for about 30 seconds.

3. Use a match to light the end of a wooden splint. Allow the fire to burn a few seconds, then blow out the flames, leaving glowing embers.

4. Check for the presence of oxygen by inserting the glowing (not burning) splint deep into the test tube. Observe if the splint bursts into flames. Remove the splint after a few seconds so the test tube does not fill with smoke.

5. Add a small amount (the size of 3 or 4 rice grains) of manganese dioxide to the test tube. Loosely stopper the test tube with a one-hole stopper for about 30 seconds. Note any temperature change. Test for oxygen with a glowing splint.

6. Clean-up: Fill the test tube with water, then slowly pour the liquid into the sink. Try to keep the solid manganese dioxide in the test tube. Again fill the test tube with water and discard the water. Transfer the solid manganese dioxide to the "Waste Manganese Dioxide" container.

III. CHEMICAL EQUILIBRIUM AND LE CHATELIER'S PRINCIPLE

CAUTION: Acids and bases are corrosive and can severely damage tissue. Wear safety goggles while handling acids. If acid or base is spilled, flood the area with water and notify the instructor immediately. If acid or base contacts with skin, remove any clothing covering the skin and flush with large amounts of water. If acid or base contacts the eyes, flush with water for 20 minutes. Wash your hands after handling acids and bases.

CAUTION: Many compounds of chromium are suspect carcinogens. Wear gloves to handle potassium dichromate. Avoid skin contact. Wash your hands after use.

A. Potassium Chromate–Potassium Dichromate Equilibrium

1. Pour 3 mL of 0.1 M potassium chromate, K_2CrO_4, solution into a clean test tube.

2. Add 1 mL of 6 M hydrochloric acid (**CAUTION!**). Mix well with a clean glass stirring rod. Note any color changes. If the mixture is green, you have contaminated the mixture with hydrogen peroxide from a graduated cylinder, test tube, or stirring rod.

3. Add 2 mL of 6 M sodium hydroxide, NaOH, solution to the test tube. **CAUTION**, handle sodium hydroxide with care. Wash your hands. Sodium hydroxide is a base which neutralizes (removes) hydrochloric acid from the equilibrium mixture. Note any color changes.

4. Clean-up: Pour the contents of the test tubes in the "Waste Chromium Salts" container or as directed by your instructor. Wash your hands.

B. Iron(III) Chloride–Iron(III) Thiocyanate Equilibrium

1. Note the color of the 0.1 M iron(III) chloride, $FeCl_3$, solution and of the 0.1 M ammonium thiocyanate, NH_4CNS, solution.

2. Pour 50 mL of deionized water into a 150-mL beaker. Add 1 mL of 0.1 M $FeCl_3$ solution and 1 mL of 0.1 M NH_4CNS solution. Stir the mixture; observe the color of the solution.

3. Pour 10 mL of the solution from Step 2 into each of four clean test tubes. Number the test tubes 1 through 4. Keep test tube No. 1 as a reference for color comparison.

4. To test tube No. 2, add 2 mL of 0.1 M $FeCl_3$. Compare the color to test tube No. 1.

5. To test tube No. 3, add 2 mL of 0.1 M NH_4CNS. Compare the color to test tube No. 1.

6. To test tube No. 4, add 2 mL of 0.1 M NH_4Cl. Compare the color to test tube No. 1.

7. Clean-up: Pour the contents of the beaker and test tubes in the "Waste Iron Salts" container or as directed by your instructor.

IV. VERIFYING A BALANCED CHEMICAL EQUATION (Group Project or Instructor's Demonstration)

1. Label nine 16 $\times$ 125-mm test tubes from 1 to 9. The test tubes must be identical in diameter.

2. Review Fundamental Laboratory Operations, Part 3a, on pages 11–15 for the proper method of using a buret.

3. Using a 25-mL buret, add the volumes of 0.5 M barium chloride solution indicated in the following table.

Test Tube No.	*Volume of 0.5 M $BaCl_2$ (mL)*	*Volume of 0.5 M K_2CrO_4 (mL)*
1	1	9
2	2	8
3	3	7
4	4	6
5	5	5
6	6	4
7	7	3
8	8	2
9	9	1

4. Warm all nine test tubes in a hot water solution for a few minutes. This will help coagulate the precipitate and avoid formation of a suspension.

5. Use a buret to add the volumes of K_2CrO_4 indicated in the table and shake each of the test tubes.

6. Allow the test tubes to remain undisturbed for 30 minutes, then measure the height in millimeters of the precipitate in each of the test tubes.

7. Clean-up: Pour the contents of the test tubes in the "Waste Chromium Salts" container or as directed by your instructor. Wash your hands.

V. REACTION OF A PENNY WITH HYDROCHLORIC ACID (Optional)

1. Obtain a U.S. penny with a date of 1983 or more recent. Use a file to remove the copper from the edge of the penny. Either completely remove the copper from the edge or remove the copper from 3 or 4 small places around the edge. Better yet, try one each way.

2. Use a pencil to write your name on the white marking spot of a 100-mL beaker, or use a stick-on label.

3. Place the penny or pennies into the beaker and pour 20 mL of 6 M hydrochloric acid (HCl) into the beaker. Make sure the pennies are covered with acid. Wash your hands.

4. Observe the pennies. Which metal seems to be reacting?

5. Cover the beaker with a watch glass and set it in a safe place for 24 hours.

6. After 24 hours, carefully fill the beaker with tap water to dilute the acid. Carefully pour the solution down the drain. Fill the beaker with tap water a second time and again discard the liquid.

7. Examine the remains of the pennies very carefully.

Name ______________________ Section __________ Date ____________

Prelab Questions for Experiment 12

________ 1. The rate of a chemical reaction may be increased by
a. increasing the frequency of molecular collisions.
b. increasing the energy of molecular collisions.
c. lowering the activation energy of the reaction.
d. all of the above

________ 2. Increasing the temperature of a reaction
a. increases the frequency of molecular collisions.
b. increases the energy of molecular collisions.
c. increases both the frequency and energy of molecular collisions.
d. lowers the activation energy of the reaction.

________ 3. Increasing the concentration of a reactant
a. increases the frequency of molecular collisions.
b. increases the energy of molecular collisions.
c. increases both the frequency and energy of molecular collisions.
d. lowers the activation energy of the reaction.

________ 4. Increasing the surface area of a solid reactant
a. increases the frequency of molecular collisions.
b. increases the energy of molecular collisions.
c. increases both the frequency and energy of molecular collisions.
d. lowers the activation energy of the reaction.

________ 5. The presence of a reaction catalyst
a. increases the frequency of molecular collisions.
b. increases the energy of molecular collisions.
c. increases both the frequency and energy of molecular collisions.
d. lowers the activation energy of the reaction.

________ 6. Which has a greater surface area?
a. 1 kg of wood sawdust
b. one large piece of wood with a mass of 1 kg

7. The rate of a reaction usually doubles for every 10°C increase in temperature. Is heat a catalyst? Explain.

____

____

8. What precautions should be observed when handling hydrochloric acid, sodium hydroxide, and hydrogen peroxide?

____

____

____

9. Given the following equilibrium, A + B $\rightleftarrows$ C + D, what will be the effect on the equilibrium if each of the following takes place?

	Will equilibrium shift to the left or right?	*Will the concentration of these substances increase or decrease?*
a. Increase in the concentration of C		A and B
b. Decrease in the concentration of A		C and D
c. Increase in the concentration of B		C and D
d. Decrease in the concentration of D		A and B

10. What special item do you need to bring to the laboratory for Procedure V?

____

Name ______________________ Section __________ Date __________

Report for Experiment 12

I. FACTORS THAT AFFECT THE RATE OF A CHEMICAL REACTION

A. Nature of the Reactants

1. Which of the metals reacted with hydrochloric acid?

____

2. Write a complete balanced equation to illustrate the reaction that occurred.

____

B. Concentration of the Reactants

1. Complete the following table:

Solution	*Relative Rate of Gas Bubble Formation*	*Reaction Started (time of day)*	*Reaction Stopped (time of day)*	*Reaction Duration (minutes:seconds)*
6 M HCl				
1 M HCl				

2. In which concentration of hydrochloric acid is the reaction faster? ______________

3. What conclusion can you make about the effect of concentration on reaction rate?

____

____

C. Surface Area of a Solid Reactant

1. Complete the following data table:

Solution	*Relative Rate of Gas Bubble Formation*	*Reaction Started (time of day)*	*Reaction Stopped (time of day)*	*Reaction Duration (minutes:seconds)*
Zn Sheet Metal				
Powdered Zn				

2. In which of the zinc/hydrochloric acid solutions was the reaction faster?

____

3. What conclusion can you make about the effect of surface area on the reaction rate?

____

____

D. Temperature of the Reactants

1. Complete the following data table:

Solution	*Relative Rate of Gas Bubble Formation*	*Reaction Started (time of day)*	*Reaction Stopped (time of day)*	*Reaction Duration (minutes:seconds)*
Ice Water				
Room Temperature				

2. At which temperature was the reaction rate faster?

____

3. What conclusion can you make about the effect of temperature on reaction rate?

____

II. CATALYSIS

1. Data. Use (+) for a positive oxygen test and (–) for a negative oxygen test:

   a. Hydrogen peroxide ____________

   b. Hydrogen peroxide with added manganese dioxide ____________

2. What function did the manganese dioxide perform?

____

3. a. Is the decomposition of hydrogen peroxide exothermic or endothermic?

   ____

   b. How were you able to tell which it was?

   ____

c. Write a complete balanced equation illustrating the decomposition of hydrogen peroxide.

____

d. Draw a potential energy vs. progress of reaction curve in the space below for each of the two systems. Use a solid line for the decomposition of hydrogen peroxide, and a diluted line for the decomposition of hydrogen peroxide in the presence of manganese dioxide.

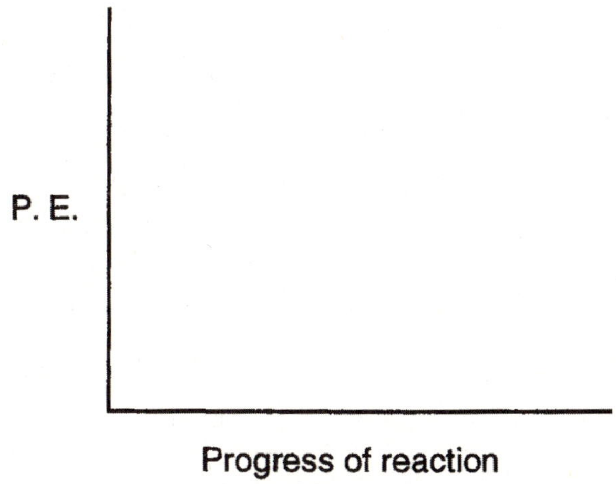

III. CHEMICAL EQUILIBRIUM AND LE CHATELIER'S PRINCIPLE

A. Potassium Chromate–Potassium Dichromate Equilibrium

1. Data. Color of Solutions:

Solution	*Color*
With hydrochloric acid	
With sodium hydroxide	

2. Write the equation for the potassium chromate/potassium dichromate equilibrium.

____

3. a. In which direction did the equilibrium shift, left or right, upon addition of HCl?

____

b. Use the equation for the equilibrium and Le Châtelier's Principle to explain the direction of the shift in equilibrium.

____

____

4. a. In which direction did the equilibrium shift, left or right, upon addition of NaOH?

____

b. Use the equation for the equilibrium and Le Châtelier's Principle to explain the direction of the shift in equilibrium.

____

____

B. Iron(III) Chloride–Iron(III) Thiocyanate Equilibrium

1. Data. Color of solutions:

Solution	*Color*
$FeCl_3$ solution	
NH4CNS solution	
Mixture of $FeCl_3$ and NH_4CNS solutions	
Test tube No. 2 (added $FeCl_3$)	
Test tube No. 3 (added NH_4CNS)	
Test tube No. 4 (added NH_4Cl)	

2. Write the equation for the iron(III) chloride/iron(III) thiocyanate equilibrium.

____

3. In which direction did the equilibrium shift, left or right, upon addition of the following reagents?

Solution	*Direction of Equilibrium Shift*
$FeCl_3$ solution (test tube No. 2)	
NH_4CNS solution (test tube No. 3)	
NH_4Cl solution (test tube No. 4)	

4. Use the equation for the equilibrium and Le Châtelier's Principle to explain the direction of each shift in equilibrium.

____

____

Name ______________________ Section ____________ Date ______________

IV. VERIFYING A BALANCED CHEMICAL EQUATION (Group Project or Instructor's Demonstration)

1. Data:

Test Tube No.	*Volume of 0.5 M $BaCl_2$ (mL)*	*Volume of 0.5 M K_2CrO_4 (mL)*	*Height of Precipitate (mm)*
1	1	9	
2	2	8	
3	3	7	
4	4	6	
5	5	5	
6	6	4	
7	7	3	
8	8	2	
9	9	1	

2. Which test tube produced the largest quantity of precipitate? ______________________

3. What molar ratio is indicated by your data?

____

4. Write a balanced equation for the reaction of potassium chromate with barium chloride.

____

5. Do the coefficients of the balanced equation agree with the molar ratio in Question 3?

____

V. REACTION OF A PENNY WITH HYDROCHLORIC ACID (Optional)

1. Which metal, copper or zinc, reacted with the hydrochloric acid?

____

2. Write a balanced equation for the reaction that occurred.

____

3. Describe the remains of the penny after 24 hours in the hydrochloric acid.

____

Name ______________________ Section __________ Date __________

Related Questions for Experiment 12

1. Parents have long told their children to thoroughly chew food before swallowing. Use what you know about the factors affecting chemical reaction rates to explain why chewed food is digested faster than unchewed food.

____

____

____

2. Use what you know about the factors affecting the rates of chemical reactions to explain why hospitals take special precautions when oxygen is administered to a patient.

____

____

____

3. Unlike today's blimps that are filled with nonflammable helium, the famous Hindenburg blimp was filled with 190,000,000 liters of hydrogen. After making 10 crossings of the Atlantic Ocean, it was destroyed by fire while landing in New Jersey in 1937, killing 36 of the 92 people on board. The cause of the fire was never determined, but it is likely that static electricity ignited leaking hydrogen. When ignited, hydrogen reacts with oxygen in the air to produce water vapor.

   a. Write a balanced equation for the burning of hydrogen.

   ____

   b. Draw a potential energy curve for the reaction.
   c. Indicate on the curve the function of the static electricity.

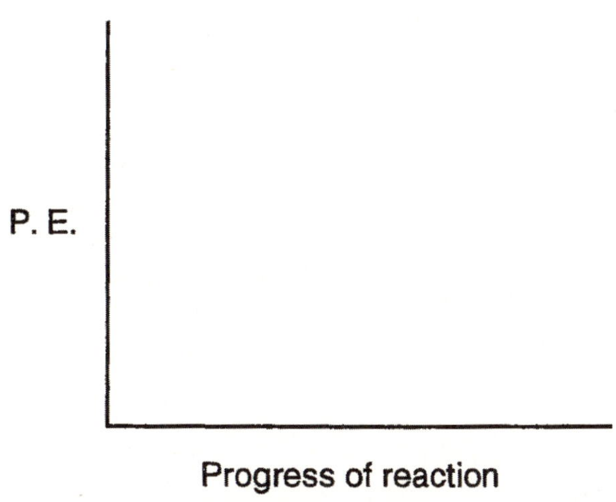

4. Commercial hydrogen peroxide is stored in dark bottles in which a trace of acetanilide has been added. Both help prevent decomposition of the hydrogen peroxide.

   a. Explain why a dark bottle is used.

   b. What function does the acetanilide perform?

   c. Draw a potential energy curve for the following two systems: a solid line to represent the decomposition of hydrogen peroxide, and a dotted line to represent the decomposition of hydrogen peroxide in the presence of acetanilide.

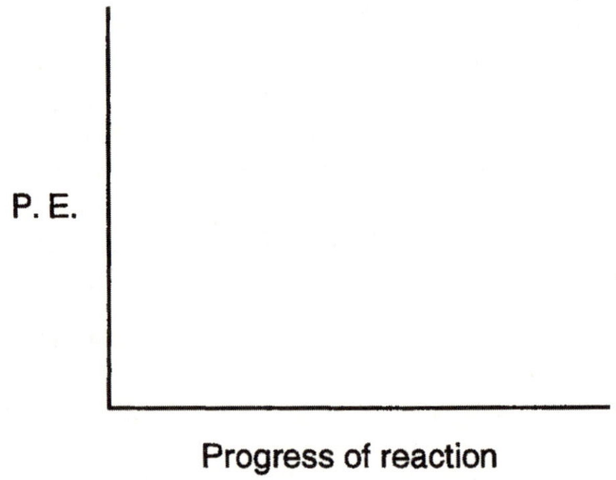

5. Give an example of stress on a chemical equilibrium in your body. Refer to your text.

Experiment 13
Antacid Analysis

Reference: *General, Organic, and Biological Chemistry: An Integrated Approach, Chapter 9.*

Purpose: To determine the relative effectiveness of commercially available antacids in reducing stomach acidity.

Introduction

HEARTBURN

The digestion of food is a mechanical and chemical process that breaks complex organic compounds into simple soluble substances absorbable by tissues. Digestion begins in the mouth and continues in the stomach and the small intestines. The digestive reactions of the stomach are catalyzed by enzymes found in the gastric juice secreted by the stomach. These enzymes are initially in an inactive form and are activated by hydrochloric acid, separately secreted into the stomach. A secreted layer of mucus protects the stomach walls from attack by the gastric juices. The mixture of food and gastric juice, called chyme, is normally prevented from backing up into the esophagus, a nine-inch-long passageway from the throat to the stomach. The esophagus is not protected by mucus and it can be painfully irritated by refluxing chyme if the esophageal muscle does not completely close. This failure can result from overeating, stress, pregnancy, or lying down soon after a meal. Fatty foods, alcohol, and chocolate adversely affect the muscle's ability to completely close; spicy foods can directly irritate the esophagus. The irritation of the esophagus by chyme results in a stinging feeling behind the breastbone. This pain is sometimes referred to as **heartburn**, **acid indigestion**, or **sour stomach**. Heartburn pain is likely when chyme of pH below 3 backs up into the esophagus.

Antacids are bases that neutralize some of the hydrochloric acid in the stomach. To be effective, an antacid should quickly raise the pH above 3 and hold it there for a period of time. Some brands of antacids will neutralize greater quantities of acid than others, some act more quickly, some are longer lasting, and some have fewer side effects. None are ideal. Most antacids include one or more of the following four bases:

Sodium bicarbonate, ordinary baking soda, is fast-acting and inexpensive. Although a good choice for occasional use, sodium bicarbonate is the least desirable for frequent use. Regular use may throw off the body's acid-base balance and encourage urinary infections and kidney disorders. People on low-sodium diets need to avoid antacids containing sodium bicarbonate. Sodium bicarbonate neutralizes stomach acid in the following manner:

$$NaHCO_3 + HCl \longrightarrow NaCl + H_2O + CO_2\,(g)$$

The buildup of carbon dioxide gas pressure may eventually lead to a "burp of relief."

Calcium carbonate (chalk) is also a potent, fast-acting, inexpensive antacid that is good for occasional use. Some brands of antacids are also promoted as calcium supplements. The maximum recommended daily intake of calcium is 3200 mg; prolonged excessive calcium intake may cause urinary tract damage, kidney stones, reduced iron levels, drug interactions, constipation, and confusion. Hydrochloric acid is neutralized by the process:

$$CaCO_3 + 2\,HCl \longrightarrow CaCl_2 + H_2O + CO_2\,(g)$$

Magnesium hydroxide, also known as milk of magnesia, is a good antacid for most people. Magnesium hydroxide has laxative effects and is often combined with calcium or aluminum compounds which are constipating. People with kidney problems must avoid long-term use of magnesium antacids—very serious problems with respiration, heartbeat, and blood pressure are possible. The neutralization reaction is:

$$Mg(OH)_2 + 2\,HCl \longrightarrow MgCl_2 + 2\,H_2O$$

Aluminum hydroxide is slow-acting, but it provides long-lasting antacid action. It is frequently combined with magnesium hydroxide to counter aluminum hydroxide's constipating properties. People with kidney disease should consult a physician before taking antacids containing aluminum. Aluminum has been accused of contributing to Alzheimer's disease but research in this area is ongoing.

$$Al(OH)_3 + 3\,HCl \longrightarrow AlCl_3 + 3\,H_2O$$

In addition to these ingredients, almost all antacids contain sweeteners, typically saccharin or sugar. Some antacids contain aspirin—the aspirin would help a headache but may cause additional stomach irritation. Other brands contain caffeine, which may stimulate additional secretion of stomach acid. Simethicone is sometimes included to reduce stomach gas but there is little evidence that this is an effective treatment. Gaviscon® antacid forms a foamy floating barrier in the stomach to prevent backup into the esophagus. Some stomach products, such as Pepto-Bismol®, are not antacids at all, and promise "soothing relief" through a "protective coating action" of bismuth subsalicylate. Three new over-the-counter products, Pepcid AC®, Tagamet HG®, and Zantac 75® are acid blockers rather than antacids; they reduce the secretion of stomach acid.

Over time, excess stomach acid can cause more than heartburn. Acid can damage the stomach lining and cause ulcers or inflammation (gastritis). Because of the side-effects of antacids and the possibility that the symptoms could be due to stomach cancer or ulcers, it is wise to see a physician before starting frequent or long-term antacid use.

EFFECTIVENESS OF ANTACIDS

Stomach acid is approximately 0.10 M hydrochloric acid. The simple and direct approach of adding 0.10 M HCl to an antacid tablet will not work because many antacids are buffered. The presence of a buffer makes it difficult to determine when all the antacid has been consumed. This problem can be avoided with an indirect analysis. You will add more hydrochloric acid than is necessary to react with the antacid. Only part of the HCl will be neutralized by the antacid.

$$HCl + \text{antacid} \longrightarrow \text{salt} + H_2O$$

The quantity of HCl not neutralized by antacid can be determined by neutralizing the excess HCl with sodium hydroxide (NaOH).

$$HCl + NaOH \longrightarrow NaCl + H_2O$$

The process of determining the amount of a solution required to react with a given amount of a sample is called a **titration**. In this experiment, a **back titration** is used to determine the amount of HCl that was not neutralized by the antacid. Bromphenol blue, an indicator that is yellow below pH 3, blue above pH 3, and green at the desired endpoint (pH 3), is used in this experiment to determine when the HCl solution has been neutralized.

Consider the following data:

Mass of antacid	5.00 g
Recommended dose	2 tablets
Volume of HCl added to 1 tablet	50.0 mL
Volume of NaOH used in titration	10.00 mL

A total of 50 mL of 0.10 M HCl was added to the antacid, some of it reacted with the antacid and some was left over, unreacted. The amount of excess, unreacted HCl was determined by titrating with 0.10 M NaOH solution. Since 1 mole of NaOH neutralizes 1 mole of HCl and since the 0.10 M NaOH has the same molar concentration as the 0.10 M HCl, every milliliter of NaOH will neutralize 1 milliliter of HCl.

$$\left(\begin{array}{l}\textbf{volume of HCl}\\\textbf{neutralized by NaOH}\end{array}\right) = \left(\begin{array}{l}\textbf{volume of NaOH}\\\textbf{used in titration}\end{array}\right)$$

If 50 mL of HCl was added to the antacid and 10 mL was neutralized by the NaOH, how much HCl was neutralized by the antacid?

$$\left(\begin{array}{l}\textbf{volume of HCl neutralized}\\\textbf{by 1 tablet of antacid}\end{array}\right) = \left(\begin{array}{l}\textbf{volume of HCl}\\\textbf{added to sample}\end{array}\right) - \left(\begin{array}{l}\textbf{volume of HCl}\\\textbf{neutralized by NaOH}\end{array}\right)$$

volume of HCl neutralized by 1 tablet of antacid = 50 mL - 10 mL = 40 mL

The recommended dosage of antacids varies by brand; the dose for the tested brand is two tablets. If 40 mL of HCl was neutralized by 1 antacid tablet, what volume of HCl was neutralized by 1 dose of antacid? This is sometimes referred to as the **Acid Neutralizing Capacity (ANC)** per dose.

$$\textbf{ANC} = \left(\begin{array}{l}\textbf{volume of HCl}\\\textbf{neutralized by}\\\textbf{1 dose of antacid}\end{array}\right) = \left(\begin{array}{l}\textbf{volume of HCl}\\\textbf{neutralized by}\\\textbf{1 tablet of antacid}\end{array}\right) \times \left(\begin{array}{l}\textbf{number of}\\\textbf{tablets}\\\textbf{per dose}\end{array}\right)$$

$$\text{ANC} = \begin{array}{l}\text{volume of HCl neutralized}\\\text{by 1 dose of antacid}\end{array} = \frac{\text{40 mL HCl neutralized}}{\text{1 tablet}} \times \frac{\text{2 tablets}}{\text{1 dose}}$$

$$\text{ANC} = \begin{array}{l}\text{volume of HCl neutralized}\\\text{by 1 dose of antacid}\end{array} = \frac{\text{80 mL HCl neutralized}}{\text{dose}}$$

The ANC is a method of comparing the effectiveness of various brands of antacids.

Procedure

1. Determine the mass of a sheet of weighing paper and record the mass on the Report. If the balance has an automatic taring feature, you may wish to use this feature and enter TARE or 0.00 g for the mass of the weighing paper.

2. Place an antacid tablet on the weighing paper and determine its mass. Record the brand name and the mass of the antacid. White tablets are easier to titrate than colored tablets.

3. Use a mortar and pestle to crush the antacid tablet. Transfer the crushed tablet to a 250-mL Erlenmeyer flask. Rinse the powder remaining in the mortar into the flask with 5–10 mL of deionized water.

4. From a 50-mL graduated cylinder, add exactly 50 mL of 0.10 M HCl. Swirl the flask to dissolve as much of the crushed tablet as possible.

5. Add 5–10 drops of bromphenol blue indicator to the flask. If the solution is yellow, record the amount of HCl added and proceed to Step 6. If the solution is blue, add 25-mL quantities of HCl (swirl the flask each time!) until a yellow color is obtained; record the total volume of HCl added.

6. Review Fundamental Laboratory Operation, part 3c, page 12 for the preparation and use of the buret. Clean and rinse a 25-mL buret. Drain the rinse water through the tip. Place the buret in a buret clamp.

7. Place about 40 mL of 0.10 M NaOH solution in a labeled beaker. Pour a few milliliters of the NaOH solution into the buret and discard through the tip. Fill the buret with the NaOH solution. Allow a small amount of the NaOH to drain through the tip until there is a solid column of solution extending down to the tip, completely free of air bubbles.

8. Adjust the level of the NaOH solution to the zero mark with 0.10 M NaOH. Record the initial reading of NaOH in the buret. Remember to read the bottom of the meniscus.

9. Use a sheet of paper under the flask to provide a white background. Titrate the yellow antacid mixture to a green endpoint. Swirl the flask as NaOH is added. The green color should persist for 30 seconds. If a colored antacid tablet was used, the endpoint may not be green; the endpoint for colored antacids is any change in color. Record the final reading of NaOH in the buret. Calculate the volume of NaOH used in the titration.

10. Calculate the volume of HCl neutralized by the antacid.

11. Repeat the above procedure with two more brands of antacids.

12. Clean-up: it is important that the NaOH solution be thoroughly rinsed from the buret.

Name ______________________ Section ____________ Date ______________

Prelab Questions for Experiment 13

1. a. What color must the indicator be prior to titration with NaOH solution? __________

   b. Why? ____

2. Given the following data, fill in the blank:

Brand of antacid	Brand X
Recommended dose	2 tablets
Mass of 1 antacid and weighing paper	2.50 g
Mass of weighing paper	0.20 g
Mass of 1 antacid tablet	
Volume of HCl added to sample	75.0 mL
Final reading of buret (NaOH)	14.50 mL
Initial reading of buret (NaOH)	0.00 mL
Volume of NaOH used in titration	
Volume of HCl neutralized by 1 tablet of antacid	
ANC (volume of HCl neutralized by 1 dose of antacid)	

Calculations:

Name ______________________ Section ____________ Date ______________

Report for Experiment 13

1. Data:

Remember to include units with each quantity.

	Antacid 1	*Antacid 2*	*Antacid 3*
Brand of antacid			
Recommended dose			
Cost of 1 tablet			
Cost of 1 dose			
Mass of 1 antacid and weighing paper			
Mass of weighing paper			
Mass of 1 antacid tablet			
Volume of HCl added to sample			
Final reading of buret (NaOH)			
Initial reading of buret (NaOH)			
Volume of NaOH used in titration			
Volume of HCl neutralized by 1 tablet of antacid			
ANC (volume of HCl neutralized by 1 dose of antacid)			

Calculations:

2. Ingredients:

   a. Brand Name of Antacid 1.____________________________

   Ingredients of Antacid 1.

   ____

   ____

   ____

   b. Brand Name of Antacid 2.____________________________

   Ingredients of Antacid 2.

   ____

   ____

   ____

   c. Brand Name of Antacid 3.____________________________

   Ingredients of Antacid 3.

   ____

   ____

   ____

3. Arrange the three antacids you tested in order of acid neutralizing capacity (ANC). List the most effective first.

Brand Name	*ANC*

4. In addition to acid neutralizing capacity, what other considerations are important in selecting an antacid?

   ____

   ____

Experiment 14

Organic Nitrogen Compounds

Reference: *General, Organic, and Biological Chemistry: An Integrated Approach, Chapter 10.*

Purpose: You will investigate the properties of amines and amides. You will prepare acetanilide, a typical amide commonly used in medicine. Two typical alkaloids, quinine and caffeine, will be studied; caffeine will be extracted from cola and purified.

Introduction

AMINES AND AMIDES

Amines are derivatives of ammonia, NH_3, in which one or more hydrogen atoms of ammonia have been replaced by an organic group. They are classified as primary, secondary, or tertiary, according to the number of R groups attached to the nitrogen.

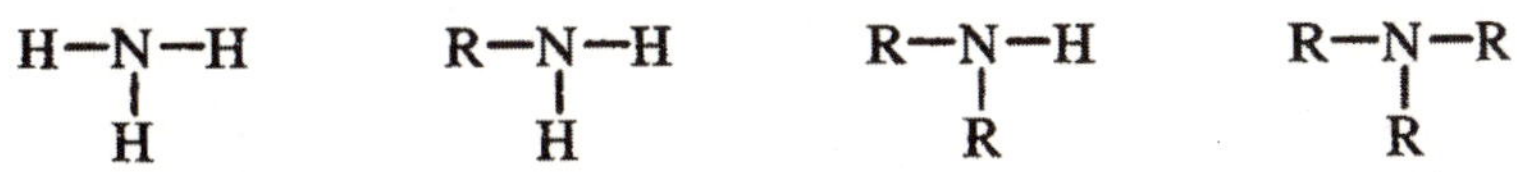

ammonia 1° amine 2° amine 3° amine

Like ammonia, amines are basic compounds that react readily with acids to form salts. Here the reaction is with hydrochloric acid.

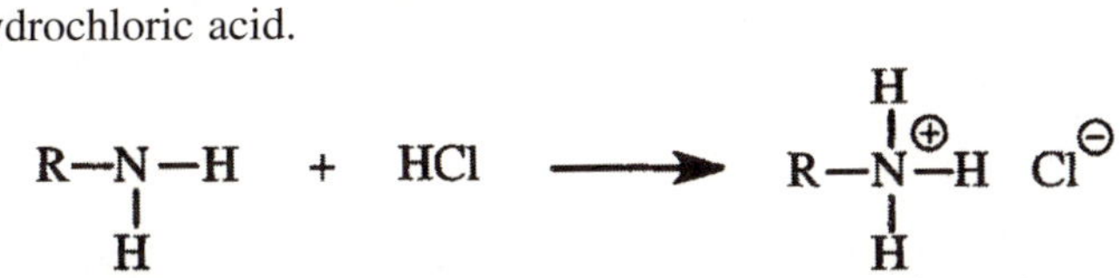

amine (base) acid amine salt

The amine may be regenerated from the salt by treatment with a base such as sodium hydroxide.

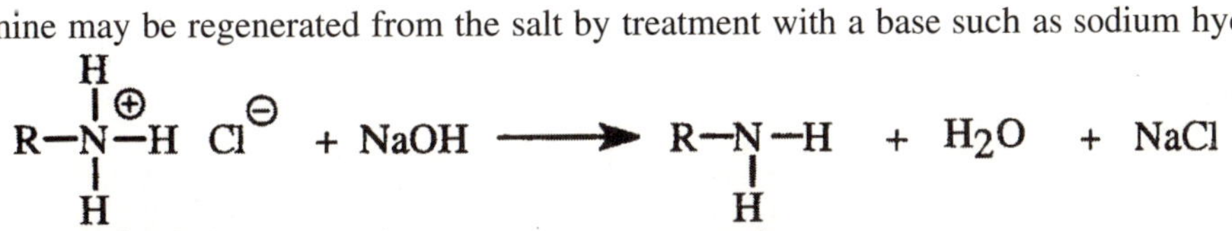

amine salt (acid) base amine (base)

You will investigate two amines in this experiment, triethylamine and aniline. Aniline is used in manufacturing dyes, medicines, plastic resins, varnishes, and perfumes.

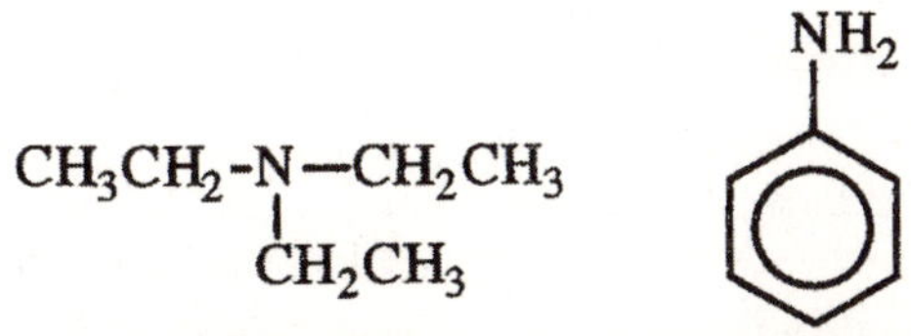

triethylamine aniline

Figure 14-1 Structural formula of amines used in Experiment 14.

Amides are ammonia or amine derivatives of carboxylic acids. They may be classified as simple (nonsubstituted), monosubstituted, or distributed as indicated in the general formulas given below.

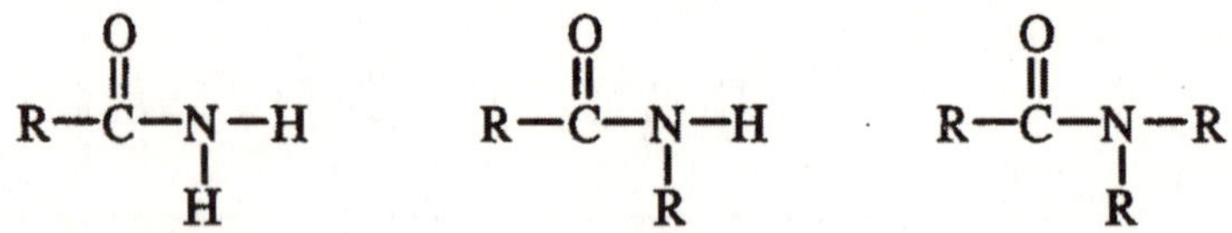

nonsubstituted amide | monosubstituted amide | disubstituted amide

The reactions of amines and amides studied in this experiment resemble reactions of proteins in body chemistry. In later studies we shall see that proteins are complex amides that are hydrolyzed in the process of digestion to amino acids—compounds containing both an amino ($-NH_2$) group and carboxyl (–COOH) group. The amino acids are absorbed into the bloodstream and transported to the cells, where they are reformed into proteins needed by the body. In summary:

$$\underset{\text{(amides)}}{\text{proteins}} + \text{water} \xrightarrow{\text{(digestion)}} \underset{\text{(amines and acids)}}{\text{amino acids}} \xrightarrow{\text{(reactions in cells)}} \underset{\text{(amides)}}{\text{body proteins}}$$

Amides can be synthesized by several methods. You will synthesize acetanilide from aniline and acetic anhydride.

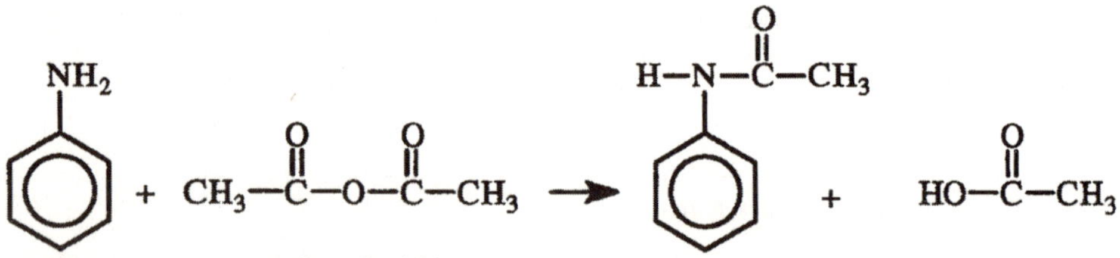

aniline | acetic anhydride | acetanilide | acetic acid

Amides, like proteins, can easily be hydrolyzed. Amides can be broken down into carboxylic acids and ammonia (or amines) as illustrated in the following equations:

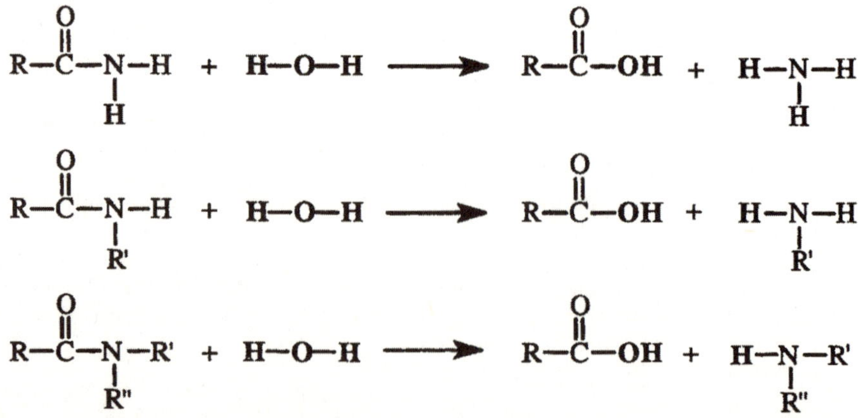

Acetamide, used in this experiment, is a solvent and is used in synthesizing many more complex organic compounds. Another amide studied in this experiment, acetanilide, is used medically as an antipyretic and analgesic; it is also used to prepare other medicinal drugs. Urea is a waste product of protein metabolism; urea is removed from the blood by the kidneys and eliminated in the urine. Urea is used as a fertilizer and to make plastic resins.

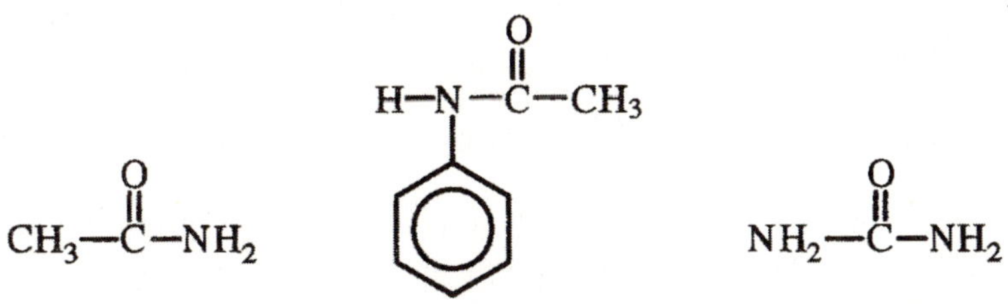

acetamide acetanilide urea

Figure 14-2 Structural formulas of amides used in Experiment 14.

ALKALOIDS

Alkaloids are naturally occurring nitrogen compounds having pronounced physiological activity. Like other simpler amines, alkaloids are bases; the name alkaloid comes from their alkaline properties. These complex molecules come from plants and many have medical uses. They generally are insoluble in water, and are often found in commercial products in the form of their water-soluble acid salts.

$$\underset{\text{(water insoluble)}}{\text{alkaloid}} + \text{acid} \longrightarrow \underset{\text{(water soluble)}}{\text{alkaloid acid salt}}$$

The pure alkaloid can be obtained as a precipitate from its acid salt by treatment with a base such as sodium hydroxide.

$$\text{alkaloid acid salt} + \text{base} \longrightarrow \text{alkaloid}\downarrow$$

Some of the better known alkaloids include atropine, used to dilate the pupils of the eyes; morphine, a painkiller; heroin, a synthetic modification of morphine that is a painkiller and narcotic; codeine, a painkiller and cough-suppressant; nicotine, a component of cigarette smoke and also a potent insecticide; reserpine, a tranquilizer; and cocaine, a local anesthetic and narcotic. The alkaloids you will study in this experiment, caffeine and quinine, were selected because of their ease of separation and their availability as components of commonly used foods.

The alkaloid **caffeine** is found in coffee, tea, and kola nuts. Caffeine is a mild stimulant to the central nervous system causing increased alertness and the ability to put off sleep. Caffeine is found in many over-the-counter medications because it tends to counteract the sedative effect of the active ingredient. The analgesic (pain-relieving) potency of aspirin may be increased by caffeine although some studies dispute this. Combination pain relievers often contain both aspirin and caffeine. Caffeine is the main ingredient of No-Doz® stay-awake tablets.

Coffee beans contain 2–5% caffeine, about the same amount present in tea leaves. Many people prefer to drink decaffeinated coffee. One of several solvents is used to extract the caffeine from the beans prior to roasting. There was once concern about the safety of the widely used solvent trichloroethylene; traces of the solvent remain in the decaffeinated coffee. The safer trichloroethane and the much safer liquid carbon dioxide are now used as solvents for the extraction.

Cola soft drink beverages are made from extract of the kola (cola) nut; the extract contains caffeine. Cola drink bottlers add phosphoric acid, caramel, sweeteners, and carbonated water. The U.S. Food and Drug Administration (FDA) requires that a "cola" contain some caffeine but limits the amount that can be present. Manufacturers usually remove all the caffeine from the kola extract and then add the correct amount. Caffeine is also added to some noncola soft drinks. Energy drinks are not included in FDA regulations and the caffeine limit is not disclosed on the labels. Some of these energy drinks have caffeine limits above the FDA limit imposed on cola or soda drinks.

Caffeine has little effect on the blood pressure in minute quantities but large amounts increase blood pressure. Individuals who drink coffee, tea, or cola drinks in large amounts can develop both a tolerance for and a dependence on caffeine. Heavy users can experience withdrawal symptoms of lethargy, headache, and even nausea after 18 hours of abstinence.

The structure of caffeine is very similar to that of adenine and guanine—important components of the genetic material DNA. This similarity has raised concern that caffeine might cause cancer or birth defects. So far there is little evidence to support this concern. Some people consider caffeine to be an addictive drug and some religions prohibit use of beverages containing caffeine.

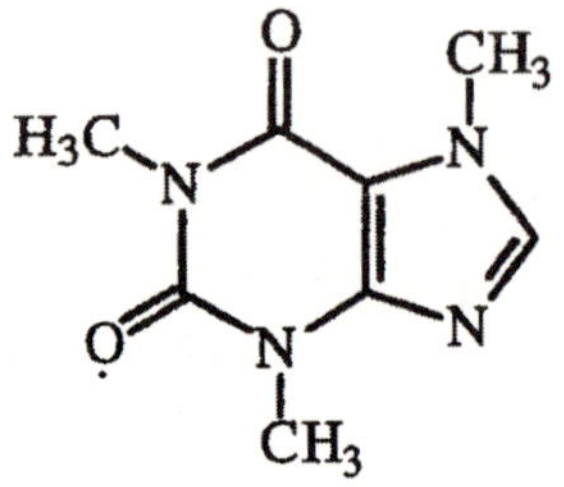

caffeine

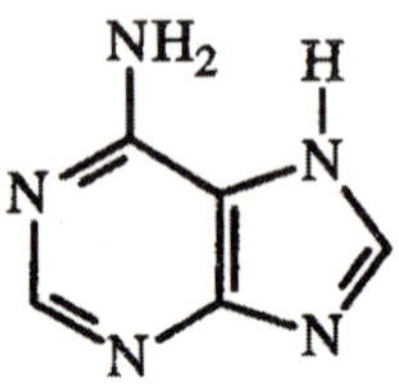

adenine

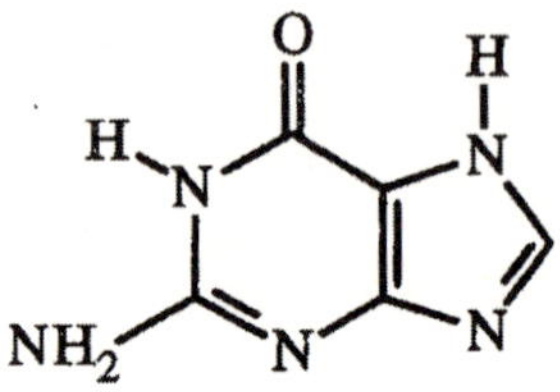

guanine

Quinine is an antipyretic (fever-reducer) and was for a long time the only known remedy for malaria. Quinine binds to the DNA of malaria-infected cells and inhibits their replication. Only infected cells are affected because they absorb quinine in higher concentrations than unaffected cells. The alkaloid is found in the bark of the cinchona tree; these trees were extensively cultivated in Indonesia in the late 19th century. When the Japanese invasion of Indonesia in World War II cut off the supply of quinine needed by Allied troops, American chemist Robert Burns Woodward succeeded in synthesizing quinine from coal tar. Woodward was renown for his ability to synthesize complex organic substances and was awarded the 1956 Nobel Prize in chemistry. Most alkaloids taste bitter, and quinine is frequently used as a standard reference for bitterness in taste studies. The bitter taste of tonic water is due to quinine. Water solutions of quinine salts are highly fluorescent, appearing light blue in the presence of ultraviolet light (blacklight).

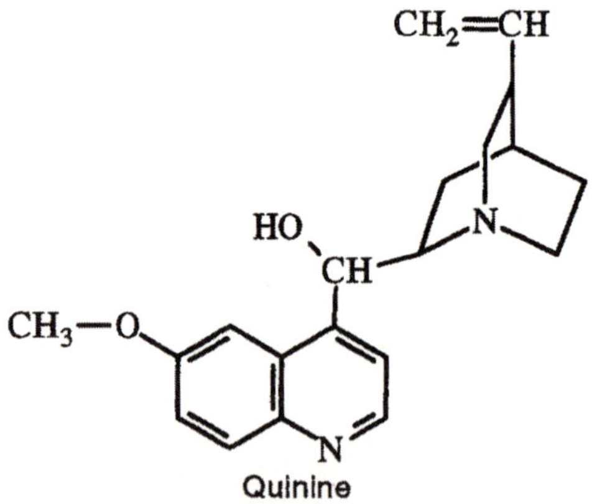

Quinine

CAUTION: Handle all organic compounds with care; avoid skin contact. Most organic compounds are flammable.

CAUTION: Triethylamine has a pungent odor.

CAUTION: Avoid skin contact with aniline, a deadly poison. Wash your hands after use.

CAUTION: Acetic anhydride produces irritation of tissues as a liquid or vapor. Avoid contact with skin or eyes. Wash your hands after use.

CAUTION: Dichloromethane is potentially harmful. Avoid inhalation and skin contact. This solvent should only be used in the hood. Wear protective gloves.

Procedure

I. BASICITY OF AMINES

CAUTION: Handle all organic compounds with care; avoid skin contact. Most organic compounds are flammable. Perform this part of the experiment in a hood. Wear gloves.

CAUTION: Triethylamine has a pungent odor.

CAUTION: Avoid skin contact with aniline, a deadly poison. Wash your hands after use.

1. Place about 5 drops of triethylamine in a test tube and add 2 mL of water. Observe the solubility of triethylamine. Test the solution with red and blue litmus paper. (Review the use of litmus paper in Experiment 11, page 187.)

2. **Cautiously** note the odor of the solution by gently fanning the vapor with your hand toward your nose. **Pregnant women should not check the odor.**

3. Add 6 M hydrochloric acid to the solution drop-by-drop until the odor disappears. Write an equation for the reaction that has occurred.

4. Repeat Steps 1 through 3 with aniline. **CAUTION: Aniline is a skin irritant. Wash your hands immediately.**

5. Disposal: Discard the solutions in the "Waste Amine" container.

II. THE PREPARATION OF ACETANILIDE

CAUTION: Perform this part of the experiment in a hood. Wear gloves.

CAUTION: Avoid skin contact with aniline, a deadly poison. Wash your hands after use.

CAUTION: Acetic anhydride produces irritation of tissues as a liquid or vapor. Avoid contact with skin or eyes. Wash your hands after use.

1. Pour about 1 mL of aniline into a 150-mL beaker.

2. Add 1 mL of acetic anhydride (**CAUTION, wash your hands**) and heat the mixture gently until it just begins to boil. Let the mixture cool.

3. Add just enough boiling water to the solid that is formed to dissolve it; then filter the hot solution. (Review Fundamental Laboratory Operations, Part 8, page 23.)

4. Set the filtrate aside to cool or place it in an ice bath if available. Proceed to Part III and complete the remainder of the experiment before continuing to Step 4.

5. Filter the crystals that have formed, spread them on a watch glass, and allow them to dry. Give your instructor the solid that has been produced. **Wash your hands**.

III. THE HYDROLYSIS OF AMIDES

1. Add about 2 mL of 6 M sodium hydroxide to a few crystals of acetamide in a test tube.

2. Warm the test tube gently, and cautiously note the odor of the gas produced. **Pregnant women should not check the odor.**

3. Place a strip of moist red litmus paper over the opening of the test tube. Identify the gas liberated.

4. Repeat Steps 1 through 3, substituting urea for acetamide.

5. Disposal: Discard the solutions in the "Waste Amine" container.

IV. PROPERTIES OF QUININE

1. Pour two 25-mL portions of tonic water into two 150-mL beakers. Place the beakers under blacklight (UV light) and observe if the samples are fluorescent.

2. Remove the blacklight and add 15 mL of deionized water to one of the tonic solutions. Place it aside for reference in Step 5.

3. Drop a piece of red litmus paper into the other portion of tonic water and add, while stirring, 10% sodium hydroxide solution drop-by-drop until the litmus turns blue. Place the beaker under blacklight to observe the amount of fluorescence.

4. Remove the blacklight and add 15 mL of 0.1 M sulfuric acid. Stir the solution.

5. Place both beakers under the blacklight and compare their appearance.

6. Dispose of the contents of the beakers as directed by your instructor.

V. EXTRACTION OF CAFFEINE FROM COLA (Optional Group Experiment)

DO THIS PART UNDER A FUME HOOD!

CAUTION: Dichloromethane is potentially harmful. Avoid inhalation and skin contact. This solvent should only be used in the hood. Wear protective gloves.

1. Pour 200 mL of cola into a 500-mL Erlenmeyer flask.

2. Add sodium carbonate until the solution is basic. (The sodium carbonate is added to remove any benzoic acid that may be present.)

3. Under a fume hood, add 25 mL of chloroform (**CAUTION**) and swirl the flask gently for at least 5 minutes.

4. Transfer the mixture to a 250-mL separatory funnel and drain the bottom layer (dichloromethane layer) into a 150-mL beaker.

5. Evaporate the solution to dryness by placing the beaker on a hot plate under a fume hood. The residue that remains is primarily caffeine.

6. The caffeine may be purified by **sublimation**—the process of changing from a solid directly to the gaseous state. The vapor produced will then solidify on a cold surface. Place an evaporating dish filled with cold water on the beaker and heat the beaker on low heat on a hotplate. The crystals that form on the underside of the evaporating dish are purified caffeine.

7. Dispose of the contents of the beakers as directed by your instructor.

Name ______________________ Section ____________ Date ______________

Prelab Questions for Experiment 14

_______ 1. What substances turn red litmus paper to a blue color?
   a. acids
   b. bases
   c. alcohols
   d. esters

2. Proteins belong to what class of organic compounds?

____

3. What are the hydrolysis products of a substituted amide?

____

4. If an amide produced ammonia upon hydrolysis, what would you conclude regarding the structure of the amide?

____

5. How does the preparation of acetanilide in Part II resemble the synthesis of body proteins?

____

6. List three physiological effects resulting from excessive intake of caffeine.

____

____

____

Name ______________________ Section ____________ Date ______________

Report for Experiment 14

I. BASICITY OF AMINES

1. Data:

Substance	*Solubility in Water*	*Test with Litmus*	*Odor*
Triethylamine			
Aniline			

2. Using structural formulas, write the equation for:

   a. the reaction of triethylamine and hydrochloric acid

   b. the reaction of aniline and hydrochloric acid

3. a. Which was more soluble in water, aniline or the product formed after addition of hydrochloric acid?

   ____

   b. Explain the reason for this result.

   ____

   ____

4. Explain why the odor of the amine disappeared as hydrochloric acid was added.

   ____

   ____

II. THE PREPARATION OF ACETANILIDE

1. Using structural formulas, write an equation illustrating the preparation of acetanilide.

2. How could you prove that the crystals formed are acetanilide?

____

____

III. THE HYDROLYSIS OF AMIDES

1. Using structural formulas, write equations for:

   a. the hydrolysis reaction of acetamide and water

   b. the hydrolysis reaction of urea and water

2. Why did the red litmus paper turn blue when acetamide and urea were hydrolyzed?

____

____

Name ______________________ Section ____________ Date ______________

IV. PROPERTIES OF QUININE

1. What compound caused the tonic water to be fluorescent?

   ____

2. Why did the fluorescence diminish after the addition of sodium hydroxide solution?

   ____

   ____

3. What reaction occurred when sulfuric acid was added in Step 4?

   ____

4. a. How did the two solutions in Step 5 compare in fluorescence?

      ____

   b. Explain the reason for this occurrence.

      ____

      ____

V. EXTRACTION OF CAFFEINE FROM COLA (Optional Group Experiment)

1. Cola tested ____

2. Describe the odor of caffeine.

   ____

3. a. Compare the quantity of caffeine isolated by your group to the quantities isolated by other groups using different colas. Which colas appear to have the most caffeine?

      ____

      ____

   b. The least caffeine?

      ____

      ____

Name ______________________ Section ____________ Date ____________

Related Questions for Experiment 14

1. The analgesic property of acetanilide was discovered by accident when it was added by mistake to a patient's prescription. Acetanilide can be toxic; therefore scientists searched for more effective, less toxic variations. Two of the compounds tested, acetaminophen and phenacetin, were found to have effective analgesic and anti-pyretic properties but only aspirin reduced inflammation. Phenacetin was once an ingredient in APC (aspirin, phenacetin, caffeine) tablets but its use has diminished because of concerns that it may cause severe kidney problems. Unlike aspirin, acetaminophen does not cause stomach irritation but overdoses can cause permanent liver damage.

   a. Draw the structural formula of acetanilide.

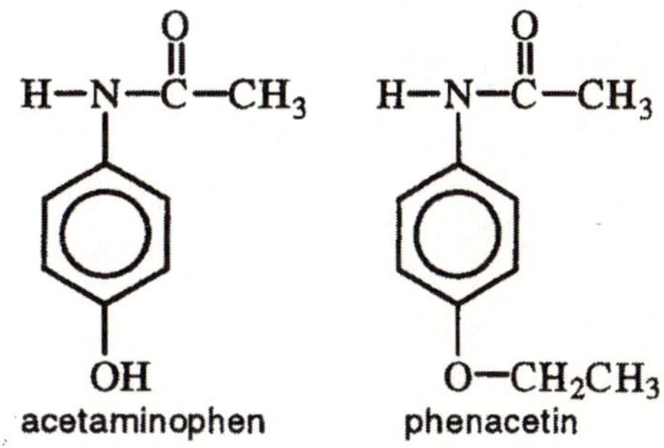

   b. What chemical changes are necessary to convert acetanilide to acetaminophen?

   ____

   c. What common over-the-counter pain relievers contain acetaminophen?

   ____

   d. What chemical changes are necessary to convert acetanilide to phenacetin?

   ____

   e. Draw the structural formula of aspirin. (See Experiment 15.)

   f. What similarities exist in the structures of aspirin, acetaminophen, and phenacetin?

   ____

2. Caffeine stimulates the cerebral cortex by inhibiting an enzyme that in turn inactivates a form of the energy-supply molecule adenosine triphosphate (ATP). ATP is one of the most important molecules in living things; it is the immediate source of biochemical energy for powering reactions such as muscle contraction. ATP is also intimately involved with the metabolism of food and the biosynthesis of protein.

a. Write the structural formula of caffeine.

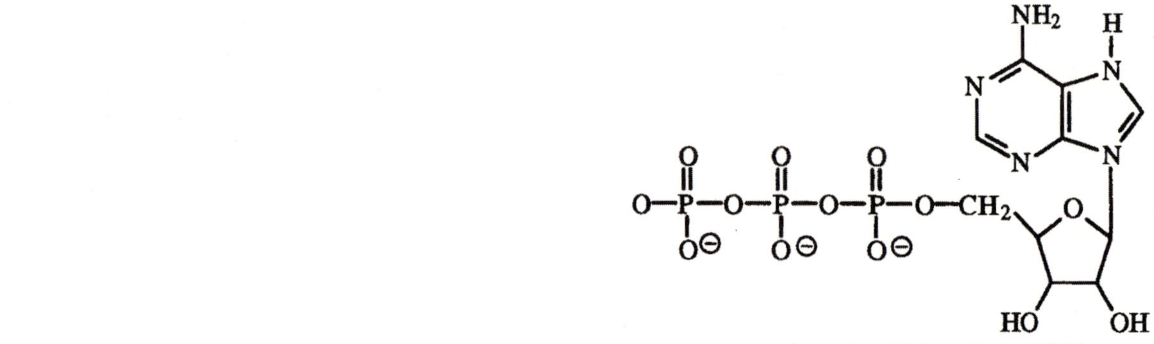

adenosine triphosphate (ATP)

b. In what way is the structure of caffeine similar to ATP?

____

____

3. The artificial sweetener aspartame (NutraSweet®) is a dipeptide—a very short protein molecule. The sweetness of aspartame was discovered accidentally in 1965 when a careless chemist licked his dirty fingers. He was fortunate that the compound was not toxic; his company was fortunate that he was curious to discover why his fingers were so sweet. Identify the functional groups (alcohol, amine, etc.) in the aspartame molecule.

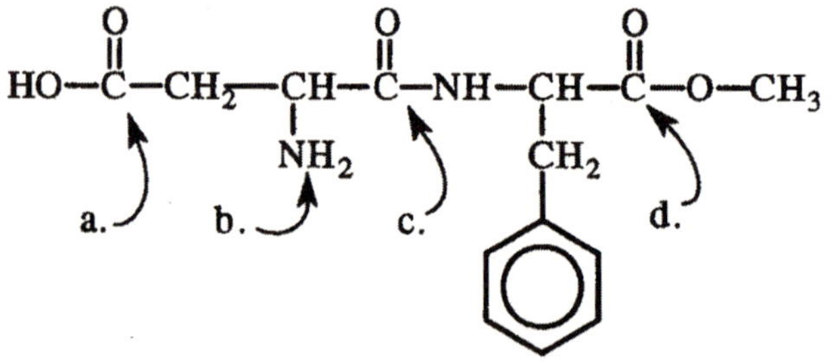

a. ____

b. ____

c. ____

d. ____

e. Aspartame decomposes slowly in liquids, so soft drinks sweetened with it have limited shelf life. Which functional group is easily hydrolyzed?

____

Experiment 15
Analysis of an Aspirin Tablet

Reference: *General, Organic, and Biological Chemistry: An Integrated Approach, Chapter 9.*

Purpose: You will examine the properties of various brands of aspirin with respect to their rate of disintegration, shelf life, composition of binder, and the percentage of aspirin in a tablet. You will hydrolyze aspirin with a strong base.

Introduction

Ancient Native Americans used the bark of the willow tree to counter fever and pain. Europeans learned of the medicinal properties of willow bark in 1763 when clergyman Edward Stone read a paper to the Royal Society of London. Willow bark extract was eventually found to be a powerful **analgesic** (pain reliever), **antipyretic** (fever reducer), and **anti-inflammatory** (reduces swelling) drug. Organic chemists were able to isolate and identify the active ingredient in willow bark in 1838. The ingredient, **salicylic acid**, was named from *salix*, Latin for the willow tree. The use of salicylic acid was limited because its acidic properties caused severe irritation of the stomach.

The German chemist Felix Hoffman, working for the firm of Bayer, synthesized an ester of salicylic acid, **acetylsalicylic acid** in 1893. Acetylsalicylic acid was marketed by Bayer under the trade name of **aspirin**. During World War I the U.S. Government seized Bayer's assets and sold them and the Bayer name to Sterling Products (now Sterling Winthrop Inc.).

Like salicylic acid, aspirin is an analgesic, antipyretic, and anti-inflammatory; it is much less irritating to the stomach than salicylic acid. Aspirin does cause slight gastrointestinal bleeding that can, over time, cause iron deficiency or gastric ulcers. These complications can be avoided with enteric-coated aspirin, which does not dissolve until reaching the small intestine.

The standard adult aspirin tablet contains 325 mg (5 grains) of aspirin—the remainder of the tablet is made up of inert materials called binders, whose function is to hold the tablet together. Some manufacturers apply a micro-coating of hydroxypropyl methylcellulose to prevent the tablet from dissolving in the mouth but permit it to dissolve in the stomach. Clinical studies among different brands of aspirin have shown no significant difference in relieving symptoms such as arthritis pain. Tablets containing aspirin sometimes include buffers to combat stomach irritation. Combination pain relievers usually contain aspirin and analgesics such as phenacetin and caffeine.

Aspirin is the most used medicine in the U.S.—more than 100 billion tablets are consumed each year for relief of headache, toothache, pain and fever of colds, muscle aches, menstrual pain, and pain of arthritis. Studies have suggested that one baby aspirin or one low dose aspirin tablet per day (both contain 81 mg of acetylsalicylic acid) may reduce the risk of heart attack, stroke, and colon cancer. Aspirin should not be given to children who have influenza or chicken pox because of the risk of the rare and often fatal Reye's Syndrome; children suffering from these diseases should be given a nonaspirin pain reliever such as acetaminophen. Unless directed by a physician, aspirin should not be taken during the last 3 months of pregnancy.

Only in the last few years have biochemists begun to understand how aspirin works its wonders. Aspirin prevents the formation of prostaglandins, a class of compounds responsible for evoking pain, fever, and local inflammation. Like many medicinal drugs, aspirin was developed from a naturally occurring substance. Chemists first isolated the active ingredient, determined its structure, and then improved on the original. Further improvement is possible once the exact mechanism of aspirin's interaction with prostaglandins is understood.

Aspirin is easily prepared in the laboratory by treating salicylic acid with either acetic anhydride or acetic acid.

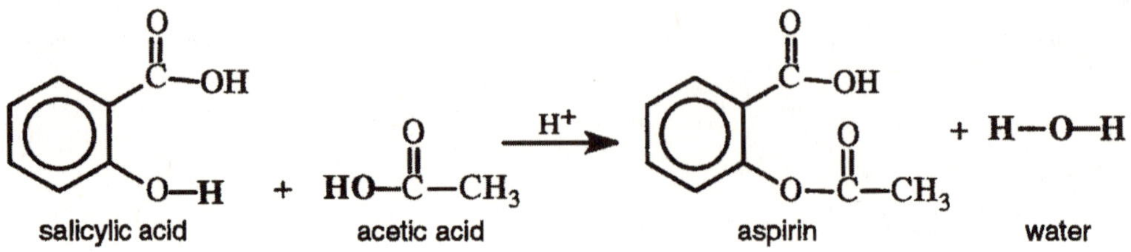

Aspirin may be reconverted into salicylic acid in either of two ways. First, it may be hydrolyzed slowly in an acidic solution as illustrated in the following equation.

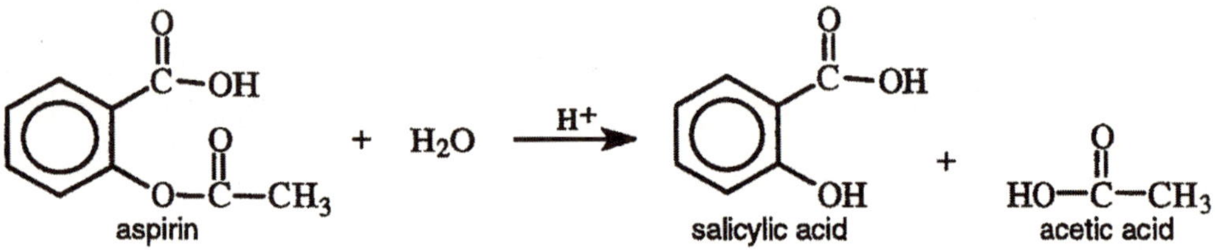

If aspirin is stored over a long period of time, especially under moist conditions, the above hydrolysis reaction takes place, and the characteristic odor of vinegar (acetic acid) is easily detected. Generally, "old" aspirin tablets contain traces of acetic acid and salicylic acid.

A second, and faster, method for converting aspirin into salicylic acid uses hydrolysis with sodium hydroxide (called saponification), followed by acidification of the sodium salt.

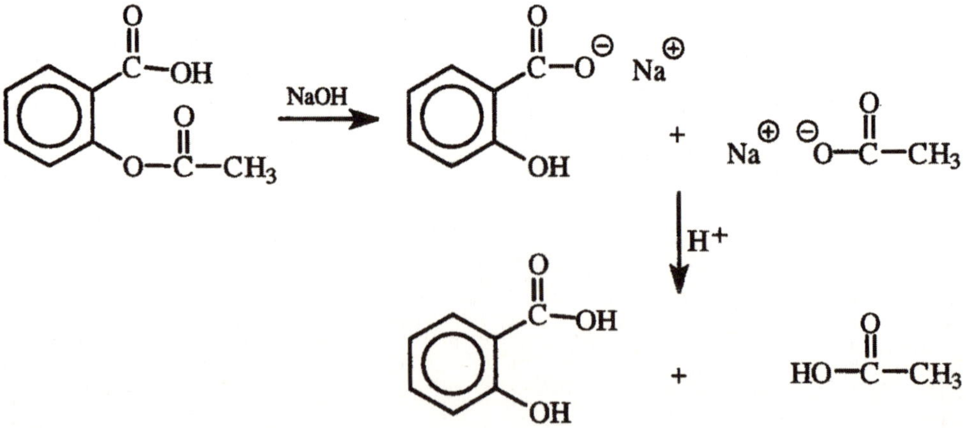

Salicylic acid can be readily detected by the presence of a violet color upon addition of iron(III) chloride, and the presence of salicylic acid in an aspirin tablet is an indication of its decomposition.

> **CAUTION: Methanol is flammable. Pregnant women may be advised by their physicians to avoid breathing vapor of evaporated methanol.**
>
> **CAUTION: Hydrochloric acid and sodium hydroxide can damage the skin; wash your hands after use.**

Procedure

I. RATE OF DISINTEGRATION

The results you obtain in this part will apply to the remainder of the experiment.

1. Simultaneously drop several different brands of aspirin into 150-mL beakers that contain 20 mL of 0.1 M hydrochloric acid.
2. The pH (acidity) of the hydrochloric acid approximates the pH of stomach acid. Observe the rate of disintegration of the tablets.

II. TEST FOR SALICYLIC ACID

1. Into one test tube, place a pinch of pure acetylsalicylic acid.
2. Into a second test tube, place a pinch of salicylic acid.
3. Add a few drops of 0.2 M iron(III) chloride, $FeCl_3$, to each of the two test tubes.
4. Dispose of the contents of the test tube as directed by your instructor.

III. STABILITY AND SHELF LIFE

1. Obtain an "old" and a fresh aspirin tablet from your instructor. Check each tablet for a vinegar odor.
2. Place each tablet in a separate clean, dry test tube and grind it to a fine powder with a glass stirring rod. Be especially careful not to contaminate one tablet with another.
3. Add several drops of 0.2 M iron(III) chloride to each of the two powders. Note the color produced **immediately** after the reagent is added.
4. Place a second fresh aspirin tablet in a clean, dry test tube and gently heat over a Bunsen burner flame until it darkens slightly. (See Fundamental Laboratory Operations, Section 6a, page 19 for the method of heating a test tube.) Cautiously smell the vapors produced.
5. Allow the test tube to cool and test for the presence of salicylic acid by adding several drops of iron(III) chloride solution.
6. Dispose of the contents of the test tube as directed by your instructor.

IV. ESTIMATING THE PERCENT OF ASPIRIN IN A TABLET

CAUTION: Methanol is flammable. Pregnant women may be advised by their physicians to avoid breathing vapor of evaporated methanol.

1. Measure the mass of a 150-mL beaker.

2. Add 3 fresh aspirin tablets to the beaker; find the combined mass of beaker and tablets.

3. Pour 20 mL of methanol in the beaker. **CAUTION: Methanol is flammable.**

4. Allow the aspirin to slowly disintegrate; break up the last few lumps with a stirring rod. The aspirin will dissolve in methanol, the binder will not.

5. Determine the mass of a clean, dry evaporating dish.

6. Prepare a filtration setup with the weighted evaporating dish as the receiving vessel. (See Fundamental Laboratory Operations, Section 8, page 23.) Stir the aspirin suspension vigorously and immediately filter it to remove the binder. Collect the filtrate in the weighed evaporating dish.

7. Pour an additional 5 mL of methanol in the beaker. Swirl the beaker to remove as much of the remaining aspirin as possible and quickly pour the suspension through the filtration setup of Step 7. Collect the filtrate in the same weighed evaporating dish. Save the residue on the filter paper for use in Part V.

8. If possible, perform this step in a fume hood. Carefully evaporate the filtrate to dryness by placing the evaporating dish over a hot-water bath. Avoid spattering! (See Fundamental Laboratory Operations, Section 5b.) **CAUTION: Methanol is flammable.**

9. Allow the evaporating dish to cool, wipe off any moisture on the bottom of the dish, and measure its mass. Record the mass of aspirin. Save the aspirin for use in Part VI.

V. COMPOSITION OF THE BINDER

1. Transfer the residue on the filter paper to a test tube, add 5 mL of water; stopper and shake the tube.

2. Add a drop of 1% iodine solution.

3. Repeat Steps 1 and 2 with a pinch of soluble starch instead of the residue.

4. Dispose of the contents of the test tubes as directed by your instructor.

VI. THE SAPONIFICATION OF ASPIRIN

CAUTION: Hydrochloric acid and sodium hydroxide can damage the skin; wash your hands after use.

1. Carefully scrape the aspirin from Part IV into a 150-mL beaker.

2. Add 10 mL of 1 M NaOH, and carefully heat the solution over an open flame or use a hotplate with a stir bar until it boils gently. Continue the gentle heating for approximately 10 minutes. Do not evaporate the solution to dryness. **CAUTION: This procedure should be done in a fume hood and gloves should be worn.**

3. Allow the beaker to cool, add a strip of blue litmus, and add 1 M HCl drop by drop (while stirring) until the blue litmus turns red.

4. Add a few drops of 0.2 M iron(III) chloride to test for salicylic acid.

5. Dispose of the contents of the beaker as directed by your instructor.

If time permits, repeat Parts IV through V with different brands of aspirin.

Name ________________________________ Section ____________ Date ______________

Prelab Questions for Experiment 15

________ 1. The presence of salicylic acid in an aspirin tablet indicates
   a. the aspirin is fresh.
   b. the aspirin is old.
   c. the aspirin contains large quantities of binder.
   d. the aspirin contains small quantities of binder.

________ 2. A sharp vinegar odor indicates
   a. the aspirin is fresh.
   b. the aspirin is old.
   c. the aspirin contains acetic acid as a binder.
   d. the aspirin has been buffered.

3. Given the following data, fill in the blank:

Mass of beaker and 3 tablets	75.42 g
Mass of beaker	74.28 g
Mass of 3 tablets	
Mass of evaporating dish and aspirin extract	42.48 g
Mass of evaporating dish	41.57 g
Mass of aspirin in 3 tablets	
Percent of aspirin in tablet	
Average mass of aspirin in 1 tablet	

Calculations:

4. a. What other chemical name is given to aspirin? ______________________________

   b. Draw the structural formula of aspirin.

   c. Label the structural formula with the names of the functional groups (aldehyde, ketone, etc.).

   d. Which functional group is responsible for causing stomach irritation?

   ____

5. a. Draw the structural formula of salicylic acid.

   b. Label the structural formula with the names of the functional groups (alcohol, ether, etc.).

   c. Which functional group reacts with acetic acid to form aspirin?

   ____

Name ________________ Section ________ Date ________

Report for Experiment 15

I. RATE OF DISINTEGRATION

1. List the brands of aspirin tested.

 ____

2. a. Which brands of aspirin seemed to break up faster? ________________

 b. Slowest? ________________

3. Do your results indicate that any one brand of aspirin would be significantly faster acting than another?

 ____

II. TEST FOR SALICYLIC ACID

Observations after addition of iron(III) chloride:

a. Pure acetylsalicylic acid ________________

b. Pure salicylic acid ________________

III. STABILITY AND SHELF LIFE

1. Data:

	Old Aspirin	*Fresh Aspirin*	*Heated Aspirin*
Odor			
$FeCl_3$ test			

2. What conclusions can you make regarding these results?

 ____

3. What precautions should you take in storing aspirin tablets over a long period of time?

 ____

4. Without using chemical reagents, how could you determine if aspirin is reasonably fresh?

 ____

IV. ESTIMATING THE PERCENT OF ASPIRIN IN A TABLET

1. Data:

Brand of aspirin			
Mass of beaker and 3 tablets			
Mass of beaker			
Mass of 3 tablets			
Mass of evaporating dish and aspirin extract			
Mass of evaporating dish			
Mass of aspirin in 3 tablets			
Percent of aspirin in tablet			
Average mass of aspirin in 1 tablet			

Calculations:

2. a. Do your experimental data indicate that the tested brands of aspirin tablet contain 5 grains of aspirin as claimed?

____

b. Why might your data show an amount different than 5 grains?

____

____

Name ______________________ Section ____________ Date ____________

V. COMPOSITION OF BINDER

1. Results of test with iodine solution:

Brand of Aspirin	*Results of Test with Iodine Solution*

Starch	

2. What conclusions can you make about the binder composition in the various brands of aspirin tablets?

____

____

VI. SAPONIFICATION OF ASPIRIN

1. Results of test with iron(III) chloride solution

____

2. What conclusion can be made concerning these results?

____

Name ______________________________ Section ____________ Date ______________

Related Questions for Experiment 15

1. In most fatal cases of aspirin poisoning, the dosage has been greater than 20 grams. How many aspirin tablets are necessary to supply 20 grams of aspirin? ________________

   Calculations:

2. Why do the textures of different brands of aspirin tablets differ?

   ____

   ____

3. Aspirin relieves the symptoms of head colds by reducing fever and easing aches and pains but does not shorten the time needed to "get over" the cold. Some physicians have suggested that aspirin may actually lengthen the period of a cold. Why might aspirin have this lengthening effect?

   ____

   ____

Experiment 16
Organic Oxygen Compounds

Reference: *General, Organic, and Biological Chemistry: An Integrated Approach, Chapter 4, Chapter 10, Chapter 11.*

Purpose: You will investigate physical and chemical properties of organic compounds containing oxygen. You will perform classification tests for aldehydes and acetone that are frequently used in medicine. Three esters will be synthesized. During the course of the experiment, you will identify the compounds you encounter as being components of commercially available products.

Introduction

The physical and chemical properties of the organic oxygen compounds are discussed in detail in your text. The simple test tube experiments you will perform illustrate some of the reactions most relevant to body chemistry. For example, the oxidation of ethanol that you will carry out in a test tube is similar in many respects to its oxidation in the liver. We shall later see that the oxidation of an aldehyde with a mild oxidizing agent is the technique used to detect the presence of sugar in the urine. In later experiments we will use a modification of the Legal test for identifying acetone in urine—the tell-tale symptom of diabetes.

ALCOHOLS

Alcohols are classified as primary, secondary, or tertiary according to the placement of the hydroxyl group on the molecule. Primary (1°) alcohols have the hydroxyl group attached to a carbon atom that is bonded to one other carbon atom.

RCH_2OH	R_2CHOH	R_3COH
1° alcohol	2° alcohol	3° alcohol

A secondary alcohol (2°) has the hydroxyl group bonded to a carbon atom that is bonded to two other carbon atoms. A tertiary alcohol has the hydroxyl group attached to a carbon atom that is bonded to three other carbon atoms.

Phenols (aromatic alcohols) are produced when one of the hydrogen atoms on a benzene ring is replaced by a hydroxyl group.

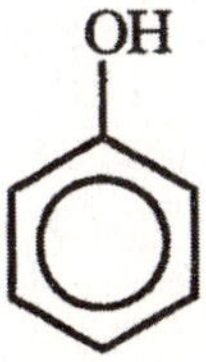

phenol

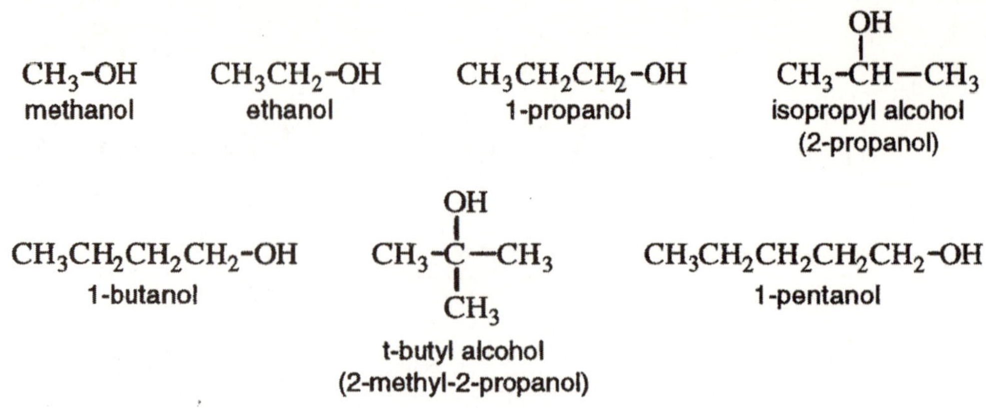

Figure 16-1 Structural formulas of alcohols used in Experiment 16.

1° methanol	1° ethanol	1° 1-propanol	2° isopropyl alcohol
1° 1-butanol	3° t-butyl alcohol	1° 1-pentanol	

Figure 16-1a Classifications of alcohols used in Experiment 16.

The hydroxyl group on an alcohol molecule forms a polar region on a nonpolar structure. The presence of the polar hydroxyl group allows alcohols with short carbon chains to be soluble in water. As the length of the chain increases, however, the nonpolar character of the carbon chain is more dominant than the polar hydroxyl group, and the alcohol's solubility in water decreases.

Phenols (aromatic alcohols) are weakly acidic. They are generally water-insoluble but dissolve in dilute base because of the formation of a soluble salt.

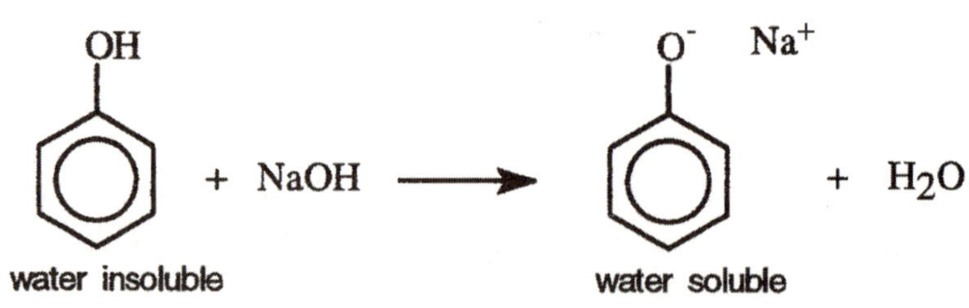

(Eq. 1)

In this experiment you will synthesize carboxylic acids (RCOOH) and ketones (R_2CO) by oxidizing a primary or secondary alcohol.

Primary alcohols can be oxidized to form aldehydes as shown in Equation 2; the intermediate aldehyde is quickly further oxidized to the carboxylic acid in Equation 3. In organic chemistry, an **oxidation** is the gain of an oxygen atom or the loss of two hydrogen atoms. The primary alcohol of Equation 2 was oxidized because it lost two hydrogen atoms. The aldehyde of Equation 3 gained an oxygen atom in its oxidation.

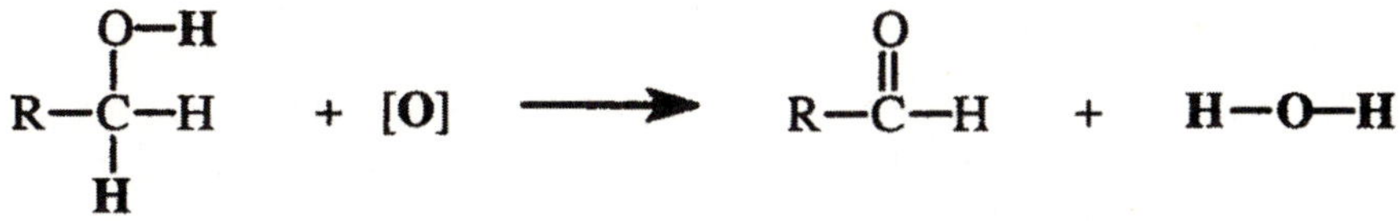

1° alcohol — oxidizing agent — aldehyde — water (Eq. 2)

aldehyde — oxidizing agent — carboxylic acid (Eq. 3)

The alcohol and aldehyde in Equations 2 and 3 are oxidized by an **oxidizing agent**, symbolized by [O]. In this experiment, the oxidizing agent is sodium dichromate ($Na_2Cr_2O_7$) in sulfuric acid (H_2SO_4). Sodium dichromate is a very strong oxidizing agent and the aldehyde formed in Equation 1 cannot be isolated; the intermediate aldehyde is immediately further oxidized to a carboxylic acid. The overall reaction with sodium dichromate and sulfuric acid is shown in Equation 4.

$$\underset{\text{1° alcohol}}{3RCH_2OH} + \underset{\text{(orange)}}{2Na_2Cr_2O_7} + 10H_2SO_4 \longrightarrow \underset{\text{carboxylic acid}}{3RCO_2H} + 11H_2O + 4NaHSO_4 + \underset{\text{(green)}}{2Cr_2(SO_4)_3} \qquad \text{(Eq. 4)}$$

When it reacts, the bright orange color of chromium(VI) in sodium dichromate is changed to the green chromium(III) ion, Cr^{3+}. Hence if a reaction takes place, there will be a sharp color change in the solution. Equation 4 is complex; we can simplify it by omitting some of the inorganic reactants and products and leaving it unbalanced (Equation 5).

$$\underset{\text{1° alcohol}}{RCH_2OH} + \underset{\text{(orange)}}{Na_2Cr_2O_7} \longrightarrow \underset{\text{carboxylic acid}}{RCO_2H} + \underset{\text{(green)}}{Cr_2(SO_4)_3} \qquad \text{(Eq. 5)}$$

Secondary alcohols react similarly to form ketones (Equations 6 and 7).

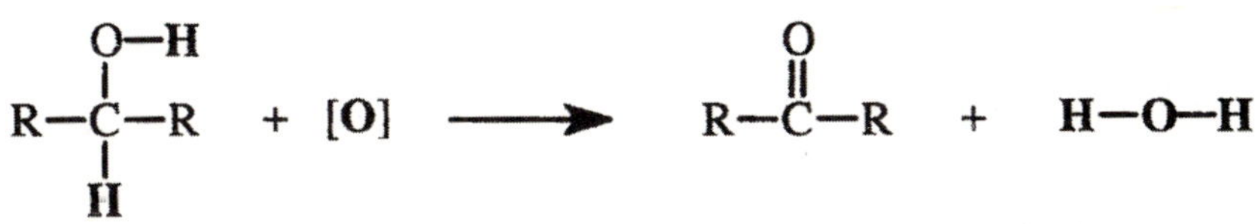

2° alcohol — oxidizing agent — ketone — water (Eq. 6)

$$\underset{\text{2° alcohol}}{R_2CHOH} + \underset{\text{(orange)}}{Na_2Cr_2O_7} \longrightarrow \underset{\text{ketone}}{RCOR} + \underset{\text{(green)}}{Cr_2(SO_4)_3} \qquad \text{(Eq. 7)}$$

Tertiary alcohols do not react and leave a clear orange solution of unreacted sodium dichromate (Equations 8 and 9).

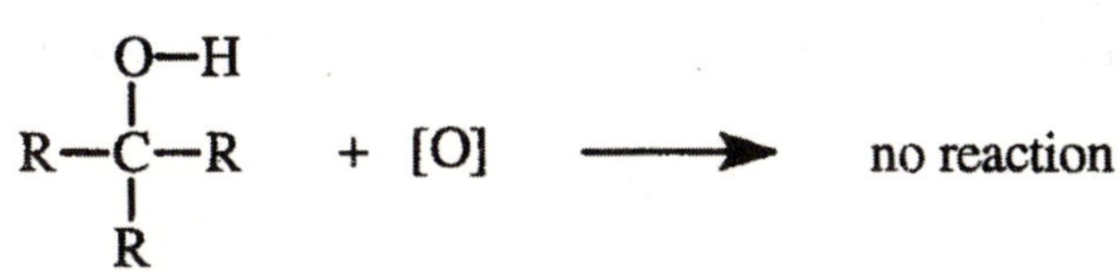

3° alcohol — oxidizing agent (Eq. 8)

$$\underset{\text{3° alcohol}}{R_3COH} + \underset{\text{(orange)}}{Na_2Cr_2O_7} \longrightarrow \text{no reaction} \qquad \text{(Eq. 9)}$$

ALDEHYDES AND KETONES

Aldehydes are easily oxidized to produce organic acids (Equations 3 and 10); ketones, however, can be oxidized only under extreme conditions. In this experiment, a dilute solution of potassium permanganate, $KMnO_4$, is used as the oxidizing agent to distinguish an aldehyde from a ketone. If a reaction occurs, the purple permanganate ion, MnO_4^-, is reduced to the brown MnO_2.

$$\underset{\text{aldehyde}}{3\,RCHO} + \underset{\text{(purple)}}{2\,KMnO_4} + H_2O \longrightarrow \underset{\text{carboxylic acid}}{3\,RCOOH} + \underset{\text{(brown)}}{MnO_2} + KOH \qquad \text{(Eq. 10)}$$

$$\underset{\text{ketone}}{R_2CO} + \underset{\text{(purple)}}{KMnO_4} \longrightarrow \text{no reaction} \qquad \text{(Eq. 11)}$$

acetone
(2-propanone)

benzaldehyde

Figure 16-2 Structural formulas of the ketone and aldehyde in Experiment 16.

THE LEGAL TEST FOR ACETONE

In cases where abnormal quantities of lipid are metabolized, as in diabetes mellitus and starvation, an increase in the level of "ketone bodies" will appear in the urine. One of the ketone bodies, acetone, can readily be detected by adding an acid solution of sodium nitroprusside to the urine specimen. The appearance of a red color is a positive test for acetone. This test is known as the "Legal test."

CARBOXYLIC ACIDS

Carboxylic acids, RCOOH, have a polar region, the carboxyl group (—COOH), on a nonpolar chain. They are, as a rule, more soluble in water than are alcohols of equal carbon length because the carboxyl group is more polar than the hydroxyl group (—OH) of the alcohol. Carboxylic acids react readily with a base to form salts, which are usually water soluble.

$$\underset{\text{carboxylic acid}}{R{-}\overset{\overset{\displaystyle O}{\|}}{C}{-}O{-}H} + NaOH \longrightarrow \underset{\text{carboxylate salt}}{R{-}\overset{\overset{\displaystyle O}{\|}}{C}{-}O^-\ N^+} + H_2O \qquad \text{(Eq. 12)}$$

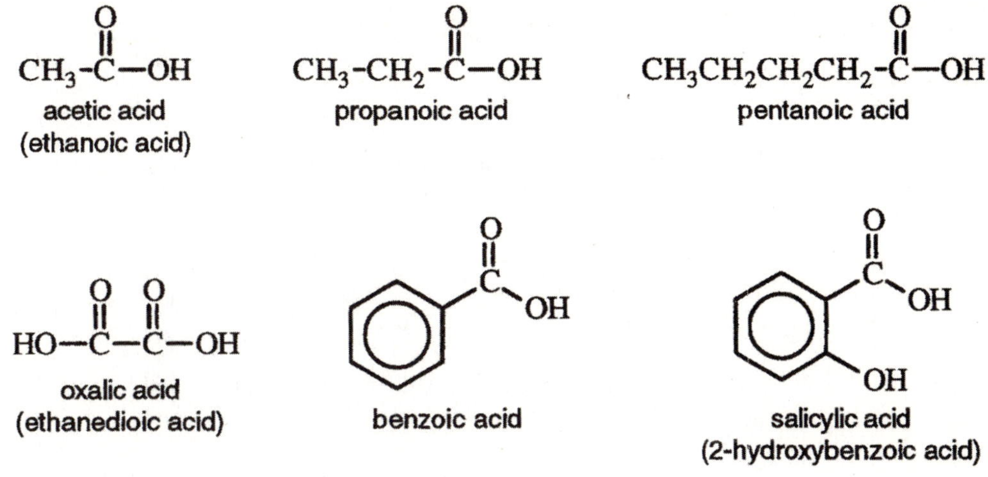

Figure 16-3 Structural formulas of carboxylic acids used in Experiment 16.

ESTERS

An ester, RCOOR, is structurally a derivative of a carboxylic acid. The hydrogen of the —OH is replaced by an alkyl (R) group, which may be different than the R of the acid. Esters are prepared by reacting a carboxylic acid with an alcohol in the presence of sulfuric acid. The acid removes water from the product side of the reaction and drives the equilibrium reaction to the right.

$$\underset{\text{carboxylic acid}}{R{-}\overset{\overset{O}{\|}}{C}{-}O{-}H} + \underset{\text{alcohol}}{H{-}O{-}R'} \xrightarrow{H_2SO_4} \underset{\text{ester}}{R{-}\overset{\overset{O}{\|}}{C}{-}O{-}R'} + \underset{\text{water}}{H{-}O{-}H} \qquad \text{(Eq. 13)}$$

Esters have pleasant odors, many of which are associated with flowers and various fruits.

CAUTION: Many compounds of chromium are suspect carcinogens; handle sodium dichromate with care. Avoid contact.

Potassium permanganate should be handled with care.

Phenol is poisonous and caustic; do not handle with bare hands.

Oxalic acid is poisonous.

Most organic compounds are flammable.

Do not smell compounds unless directed to do so; physicians may advise pregnant women to avoid smelling any compounds in this experiment.

Wash your hands after handling these and other substances.

For handling chromium (VI) compounds, potassium permanganate, phenol, and oxalic acid, gloves should be worn.

Check with your instructor on how to dispose of these chemicals properly.

Procedure

I. PROPERTIES OF ALCOHOLS

1. Place 4 drops of the following alcohols in four numbered test tubes as follows:

   No. 1: Ethanol
   No. 2: 1-Propanol
   No. 3: 1-Butanol
   No. 4: 1-Pentanol

2. Note the color and carefully check the odor of each alcohol by fanning the vapors toward your nose. **Do not perform this Step if you are pregnant.**

3. Add 3 mL of deionized water to each test tube; cork and shake vigorously. Classify the compounds as insoluble (I) or soluble (S).

4. Review the use of universal indicator, pH paper, and litmus paper in Experiment 11, page 187.

5. Test each of the water-soluble solutions (S) with wide-range pH indicator paper and record the pH.

6. Add a drop or two of universal indicator to each of the water-soluble solutions. Mix well! Do not discard the solutions until you have completed Part II.

II. PROPERTIES OF PHENOL

1. With a spatula, place one or two very small crystals of phenol into each of two test tubes. **CAUTION: Phenol is a skin irritant; avoid skin contact! Wear gloves.**

2. Cautiously note the odor. **Do not perform this Step if you are pregnant.**

3. To one test tube add 3 mL of deionized water; cork and shake vigorously.

4. Test a drop or two of the solution with red and blue litmus paper.

5. Add a drop of universal indicator to the solution. Mix well! Compare the color produced to those of the alcohol solutions of Part I, Step 5.

6. To the second test tube add 3 mL of 10% NaOH solution; cork and shake vigorously.

7. Disposal: Discard the solutions from Parts I and II into the "Waste Organic Solvent" container or as directed by your instructor. Wash the test tubes with detergent to remove water-insoluble organic compounds.

III. OXIDATION OF ALCOHOLS

1. Pour 250 mL of tap water into a 400-mL beaker and heat the water using a hot plate until it reaches a temperature of approximately 60°C. **CAUTION: use a hot plate only, not a flame, for this procedure.**

2. Use disposable laboratory gloves for this Step. Pour 2 mL of ethanol into a test tube and add 6 mL of 2% sodium dichromate. Gently shake the solution. Wash your hands.

3. Use disposable laboratory gloves for this Step. With a medicine dropper, **cautiously** add approximately 10 drops of concentrated sulfuric acid (H_2SO_4) to the test tube, gently shaking the solution after the addition of each drop. **Immediately wash your hands.**

4. Place the test tube in the warm-water bath for about 2 minutes.

5. Remove the test tube from the water bath, and note the odor produced and the color of the solution. Pregnant women should not check the odor.

6. Repeat Steps 2 through 5, substituting isopropyl alcohol for ethanol.

7. Repeat Steps 2 through 5, substituting t-butyl alcohol for ethanol.

8. Disposal: Discard the three solutions into the "Waste Chromium Salts" container or as directed by your instructor. Save the warm-water bath for Part VI.

IV. ALDEHYDES AND KETONES—CLASSIFICATION TESTS

A. Distinguishing an Aldehyde from a Ketone

1. Pour about 3 mL of benzaldehyde into a test tube and try to identify its odor in one of the commercial products your instructor has available. **Pregnant women should not check the odor.**

2. Add about 8 drops of 10% potassium permanganate solution, cork the test tube, and shake vigorously. Note the color of the solution. Wash your hands.

3. Repeat Steps 1 and 2 with acetone.

4. Disposal: Discard the two solutions into the "Waste Permanganate" container.

B. Legal Test for Acetone

1. Pour about 1 mL of acetone into a clean test tube and add 5 mL of deionized water.

2. Add a drop or two of 10% sodium hydroxide (NaOH) solution and then add several drops of a 5% solution of sodium nitroprusside. Observe the color of the solution. Wash your hands.

3. Repeat the Legal test (Steps 1 and 2) with commercial fingernail polish remover.

4. Disposal: Discard the two solutions into the "Waste Nitroprusside" container.

V. PROPERTIES OF SOME CARBOXYLIC ACIDS

1. Place the following carboxylic acids into five numbered test tubes:

No. 1:	4 drops of glacial (100%) acetic acid
No. 2:	4 drops of propanoic acid
No. 3:	4 drops of pentanoic acid
No. 4:	0.1 g oxalic acid
No. 5:	0.1 g benzoic acid

2. Note the color and carefully check the odor of each tube. **Pregnant women should not check the odor.**

3. Add 3 mL of deionized water to each test tube, cork and shake vigorously. Classify the carboxylic acids as insoluble (I) or soluble (S).

4. Test 1 mL of each of the water soluble (S) solutions with 2 drops of universal indicator solution and record the pH.

5. To the remaining 2 mL of each of the water soluble (S) solutions, add 5 mL of 15% calcium chloride ($CaCl_2$) solution, cork, and shake vigorously. Indicate with a (+) on your Report Sheet if a precipitate is formed; use a (–) if none appears.

6. To each of the insoluble (I) carboxylic acids add 3 mL of 10% sodium hydroxide (NaOH) solution; cork and shake vigorously. Use a (+) or (–) to indicate if a reaction has taken place. Wash your hands.

7. Disposal: Discard the solutions into the "Waste Organic Acid" container. Wash your hands.

VI. PREPARATION OF SOME ESTERS

1. Remove all flammable reagents from your laboratory bench. Prepare a warm-water bath as before, with temperature approximately 60°C.

2. Use disposable gloves for this Step. Pour about 2 mL of acetic acid into a test tube and add about 5 drops of concentrated sulfuric acid. **Handle concentrated sulfuric acid with caution. Immediately wash your hands.**

3. Add about 2 mL of ethanol and place the test tube in the water bath for about 10 minutes. Note the odor produced. **Pregnant women should not check the odor.**

4. Repeat Steps 1 through 3 substituting the following carboxylic acid/alcohol combinations for acetic acid and ethanol:
   1 g of salicylic acid and 3 mL of methanol
   2 mL of 1-pentanol and 2 mL of acetic acid

5. Try to identify the esters you have synthesized by comparing them with commercial products.

6. Disposal: Discard the solutions into the "Waste Organic Solvent" container.

Name ______________________________ Section ____________ Date ______________

Prelab Questions for Experiment 16

________ 1. Which of the following tests will be most helpful in diagnosing diabetes mellitus?
a. Measuring the specific gravity of the urine.
b. Treatment of urine with sodium nitroprusside reagent.
c. Heating the urine specimen.
d. Taking the pH of the urine.

2. Write structural formulas for the following compounds:

a. ethanol

b. acetone

c. acetic acid

d. isopropyl alcohol

e. methanol

3. Classify each of the following as a primary (1°), secondary (2°), or tertiary (3°) alcohol:

_______ a. $CH_3CH_2CH_2$-OH

_______ b. 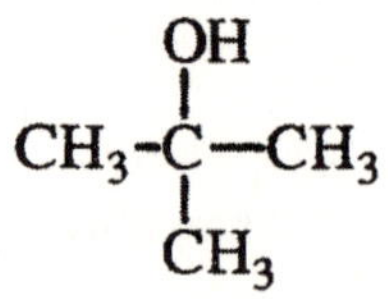

_______ c. $CH_3CH_2CH_2CH_2CH_2$-OH

_______ d. 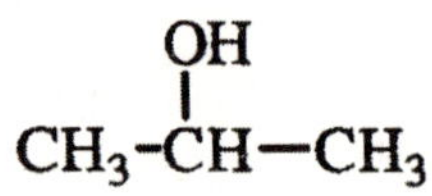

4. Water is a product in each of the following reactions. Complete each equation by writing the formula of water and the structural formula of the product.

a. $CH_3CH_2CH_2CH_2\text{-OH} \xrightarrow{[O]}$

b. 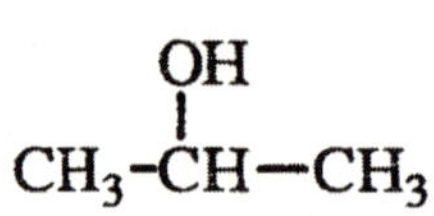 $\xrightarrow{[O]}$

c. 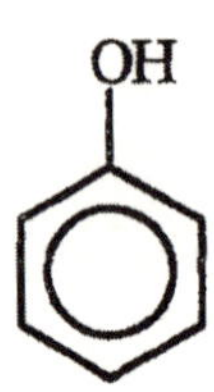 + KOH $\longrightarrow$

d. 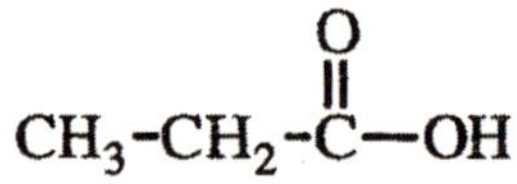 + HO—CH_3 $\xrightarrow{H_2SO_4}$

e. 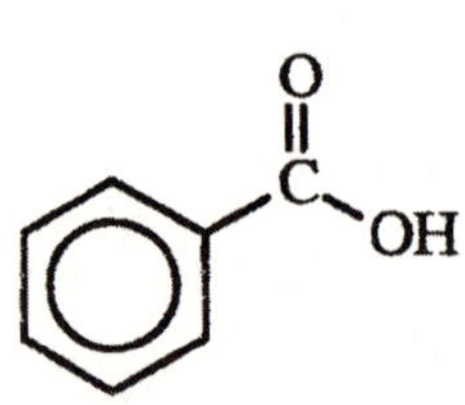 + HO—CH_2CH_3 $\xrightarrow{H_2SO_4}$

Name ______________________ Section ____________ Date ______________

Report for Experiment 16

I. PROPERTIES OF ALCOHOLS

1. Data:

Compound	*Structural Formula*	*Color*	*Odor*	*Solubility in H_2O (I or S)*	*pH*
Ethanol					
1-Propanol					
1-Butanol					
1-Pentanol					

2. a. What do your experimental results indicate concerning the relationship between the length of the carbon chain and the water solubility of alcohols?

   ____

   b. Explain the reason for this relationship.

   ____

3. Alcohols (ROH) are classified in a distinct category apart from metal hydroxides such as NaOH. Explain how your data supports different categorizations.

   ____

II. PROPERTIES OF PHENOL

1. Describe the odor of phenol. (Does its odor resemble that of any commercial product?)

   ____

2. Does phenol appear to be more soluble in water or in a 10% NaOH solution?

   ____

3. What was the pH of the phenol/water solution? ______________________

4. After adding a universal indicator, how did the phenol-water solution compare in color with the alcohol solutions in Part I?

____

5. Phenols are in some respects similar to alcohols, but phenols and alcohols are placed in separate, distinct categories. Explain how your data supports different categorizations.

____

III. OXIDATION OF ALCOHOLS

1. a. Write the structural formula for the organic product when ethanol is oxidized by sodium dichromate:

$$CH_3CH_2\text{-}OH \xrightarrow{[O]}$$

   b. What well-known fruit has an odor resembling the product of the above reaction?

____

2. a. Write the structural formula for the organic product formed when isopropyl alcohol is oxidized by sodium dichromate:

$$CH_3\text{-}\overset{\overset{\displaystyle OH}{|}}{C}H-CH_3 \xrightarrow{[O]}$$

   b. What commercial product has a similar odor? _______________

3. a. Did a reaction occur when t-butyl alcohol was heated with sodium dichromate? _____

   b. Explain the reason for this result.

____

IV. ALDEHYDES AND KETONES—CLASSIFICATION TESTS

A. Distinguishing an Aldehyde from a Ketone

1. a. Which of the commercial products has an odor similar to benzaldehyde?

____

   b. To acetone?

____

2. a. Which of the compounds tested, benzaldehyde or acetone, reacted with 10% potassium permanganate?

____

Name ______________________ Section ____________ Date ______________

b. Where a reaction has taken place, write the structural formula for the organic product formed. Write "no reaction" if there was none.

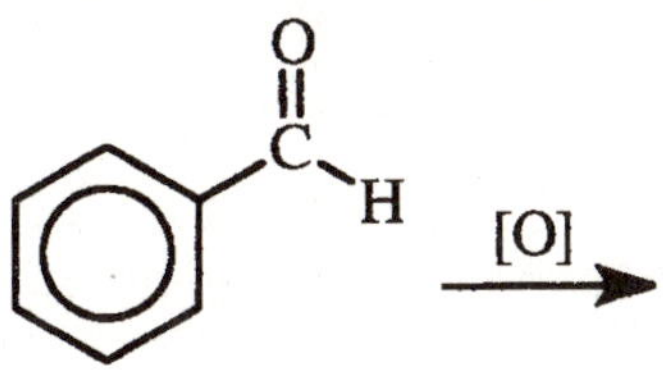

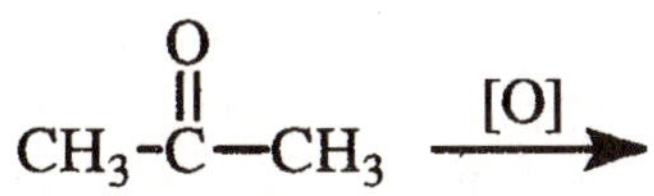

B. Legal Test for Acetone

Substance Tested	*Color after adding sodium nitroprusside*	*Acetone Present? (+ or –)*
Acetone		
Fingernail polish remover		

V. PROPERTIES OF SOME CARBOXYLIC ACIDS

1. Data

Compound	*Structural Formula*	*Solubility in Water (I or S)*	*Solubility in 10% NaOH (I or S)*	*pH*	*Reaction with $CaCl_2$ (+ or –)*
Acetic acid					
Propanoic acid					
Pentanoic acid					
Oxalic acid					
Benzoic acid					

2. How do the solubilities of acetic acid, propanoic acid, and pentanoic acid in water compare with the solubilities of alcohols of equal carbon length? Explain.

____

____

3. Explain the solubility results of benzoic acid with water and 10% NaOH.

____

____

VI. PREPARATION OF SOME ESTERS

1. Write a complete balanced equation illustrating the preparation of esters from the following carboxylic acid/alcohol combinations:

   a. Acetic acid/ethanol

   b. Salicylic acid/methanol

   c. Acetic acid/pentanol

2. Associate each of the above esters with a commercial product or fruit.

	Name of Ester	*Product or Fruit*
a.		
b.		
c.		

3. What is the function of the sulfuric acid in the above reactions?

____

Name ________________ Section ________ Date ________

Related Questions for Experiment 16

1. Compare the temperatures used in the oxidation of alcohols (Part III) with the temperature used in oxidizing an aldehyde (Part IV A). Is it easier to oxidize an alcohol or an aldehyde?

____

2. A student found an old bottle of benzaldehyde that had a poorly fitted lid. Although aldehydes are neutral compounds, the contents of the bottle turned blue litmus paper red. What has happened to the benzaldehyde? Hint: Benzaldehyde reacts easily with the air.

____

____

3. The labels have been partially washed off three bottles in the laboratory. From the labels, you know the compounds are propionaldehyde, acetone, and acetic acid. Design tests that can be run on a few drops of the chemical in each bottle to determine what compound is in each bottle. Explain the observations you would make with each test for each compound.

____

____

____

____

____

____

4. a. Write an equation for the neutralization of benzoic acid by sodium hydroxide.

   b. The organic product of the above reaction is sodium benzoate. Use *The Merck Index* to find the principal use of sodium benzoate.

   ____

5. Use *The Merck Index* to find the principal use of calcium propionate (the calcium salt of propionic acid).

____

6. a. Calcium ions are needed for blood coagulation. What were the results in Part V of treating calcium chloride with oxalic acid?

____

   b. Explain why oxalic acid is a deadly poison.

____

Experiment 17
Carbohydrates

Reference: *General, Organic, and Biological Chemistry: An Integrated Approach, Chapter 12.*

Purpose: You will perform identification tests for carbohydrates, reducing sugars, and polysaccharides. Carbohydrate digestion and absorption will be studied. Molecular models will be used to illustrate the structural differences between the α– and β– forms of D-glucose.

Introduction

Carbohydrates are a class of compounds that includes polyhydroxy aldehydes, polyhydroxy ketones, and large molecules that can be broken down to form polyhydric units that are produced upon hydrolysis. Monosaccharides cannot be broken down into smaller units. In summary:

$$\text{polysaccharide} \xrightarrow{H_2O} \text{many monosaccharide units}$$

$$\text{trisaccharide} \xrightarrow{H_2O} \text{three monosaccharide units}$$

$$\text{disaccharide} \xrightarrow{H_2O} \text{two monosaccharide units}$$

Monosaccharides, or simple sugars, are classified by the number of carbon atoms in the molecule and also by the kind of carbonyl functional group found in the structure. Thus, glucose is an aldohexose (an aldehyde containing six carbon atoms), and fructose is a ketohexose (a ketone with six carbon atoms).

DIGESTION AND ABSORPTION OF CARBOHYDRATES

Before the cells of your body can utilize the energy stored in carbohydrates present in your diet, the carbohydrates must be digested and absorbed. **Digestion** is the process by which complex molecules are broken down into simple molecules. These simple molecules pass through the intestinal wall into the bloodstream during **absorption**.

The digestion of carbohydrates begins in the mouth as teeth tear food into tiny pieces; smaller pieces have a greater surface area and will be digested faster. Saliva contains an enzyme (amylase) that begins the hydrolysis of starch (a large polysaccharide) to dextrins (small polysaccharides) and maltose (a disaccharide). After swallowing, the food enters the stomach where protein and fat digestion begin but carbohydrate digestion temporarily ceases; the low pH of the stomach's gastric juice inactivates the salivary enzymes.

As food passes into the small intestine it is neutralized by alkaline pancreatic and intestinal juices. Those juices also contain an enzyme that renews the hydrolysis of complex carbohydrates. Eventually all polysaccharides and disaccharides are broken down to glucose, fructose, and galactose. These monosaccharides are small enough in size to pass through the intestinal wall and are absorbed into the blood. After circulating in the blood, fructose and galactose are converted into glucose by the liver. The glucose in the blood may be immediately used to provide energy for cellular activities or it may be stored as glycogen (a polysaccharide) in the liver and the muscles.

starch ⟶ dextrins ⟶ maltose

sucrose ⟶ glucose + fructose

maltose ⟶ 2 glucose

lactose ⟶ glucose + galactose

Starch and cellulose are the best known polysaccharides. Ordinary table sugar is the disaccharide sucrose. The digestion of starch and sucrose will be investigated in Part V of this experiment.

Absorption is a form of **dialysis**—the movement of small molecules through a membrane. In Part I of this experiment a solution of starch and glucose will be placed inside a cellophane dialysis bag. After the dialysis bag has been immersed in water for a length of time, the water will be tested for the presence of starch and glucose.

THE STRUCTURES OF GLUCOSE

An aqueous solution of glucose is an equilibrium between three forms: α-glucose, β-glucose, and open-chain glucose. The open-chain structure makes up only 0.2 percent of the equilibrium mixture. Aqueous glucose exists 36 percent in the α structure and 64 percent in the β structure. α- and β-glucose are cyclic structures in which five carbon atoms and one oxygen atom form a six-sided ring. Six-sided glucose rings are not flat but exist in a puckered "chair" conformation. This chair structure is difficult to draw so the ring structures are often drawn flat even though this is not an accurate representation of the three-dimensional structure. The only difference between the α and β forms is the position of one hydroxyl (–OH) group on carbon-1. If the hydroxyl (-OH) group is positioned pointing down, this it the α form. If the hydroxyl group is up, it is the β-form. In Part VI of this experiment, you will use molecular models to study the structural differences of the three forms.

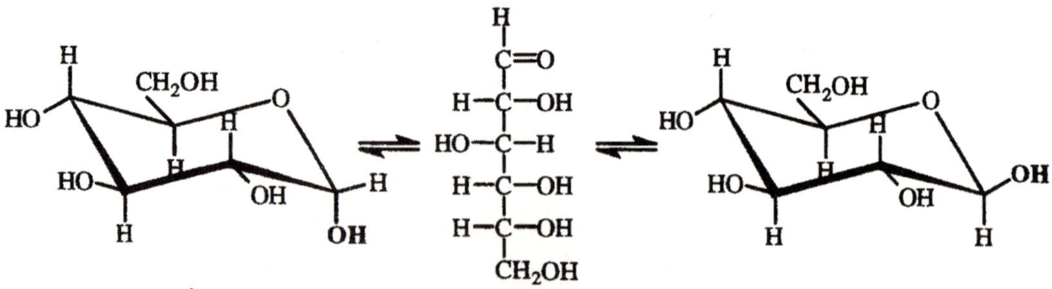

Figure 17-1 Three forms of glucose are in equilibrium in aqueous solution: the α-ring form (36%), the open-chain form (0.02%), and the β-ring form (64%). The ring forms are drawn to represent the three-dimensional shape of the structure.

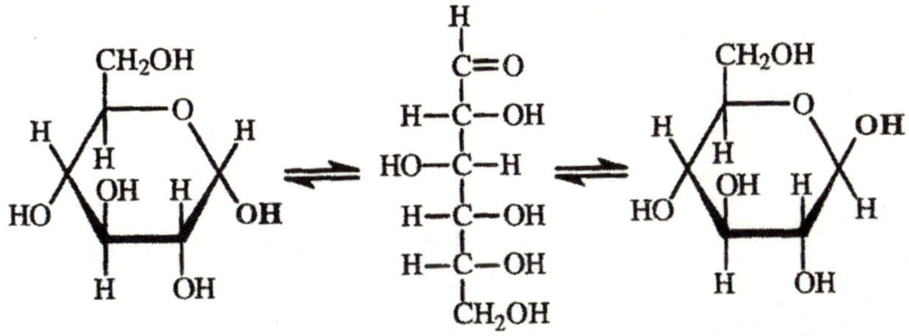

Figure 17-2 The three equilibrium forms of glucose; the α and β forms are represented by Haworth formulas that do not show accurate three-dimensional shape. α- and β-glucose contain a **hemiacetal** functional group formed when the hydroxyl group of carbon-5 of the open-chain structure reacted with the aldehyde group in the same structure.

TESTS FOR CARBOHYDRATES

A. The Molisch Test

This is a test for any carbohydrate. A small quantity of solution to be tested is poured into a test tube, and an alcoholic solution of α-naphthol is added. The tube is then held at an angle while cold, concentrated sulfuric acid is slowly poured down the side. A purplish color indicates a positive test.

B. Seliwanoff's Test

This test is used to distinguish a ketose from an aldose. The test solution is mixed with hot hydrochloric acid and resorcinol (Seliwanoff's reagent). Ketohexoses will produce a red color; aldohexoses will produce a pink color. The test is of little value when performed on a disaccharide or polysaccharide since aqueous hydrochloric acid will hydrolyze them to mixtures of monosaccharides.

C. Tests for a Reducing Sugar

These tests indicate the presence of an aldehyde group. The general reaction is:

$$\underset{\text{(blue)}}{Cu^{2+}} + \underset{\text{(free aldehyde)}}{\text{reducing sugar}} \xrightarrow{\text{heat, base}} \underset{\text{(brick red)}}{Cu_2O} + \text{oxidized sugar}$$

The reagent used is called Benedict's reagent, but many modifications, such as the Clinitest reagent, are frequently substituted. All monosaccharides will provide a positive Benedict's test, even those that do not contain an aldehyde group. Monosaccharides that are ketoses rearrange their structures to form an aldehyde group under test conditions and, therefore, give a positive test. Only those disaccharides having a free aldehyde group will react with the test reagent; sucrose is the only common disaccharide that does not react. Sugars that react are **reducing sugars***; those that are unreactive are **nonreducing sugars**.

---

*Copper(II) ion, Cu^{2+}, of Benedict's reagent is **reduced** to copper(I) ion, Cu^{+}, in the red product Cu_2O. Sugars that reduce Cu^{2+} are **reducing sugars**. The aldehyde group of the sugar gains an oxygen atom to become a carboxylic acid; the reducing sugar is **oxidized** in the course of the reaction.

D. Test for a Polysaccharide

A dilute solution of iodine will produce the following colors when mixed with a carbohydrate: starch—blue; dextrins—red; disaccharides and monosaccharides—colorless. Iodine solutions will not give an identifying color with cellulose.

> **CAUTION: Handle concentrated sulfuric acid (Part IIA, Molisch test) with care. Concentrated sulfuric acid can burn and blister the skin. Wash you hands after use. If possible, wear protective gloves.**
>
> **Seliwanoff's reagent (Part IIB) contains resorcinol, a highly irritating substance; it also contains hydrochloric acid. Wash your hands after use. If possible, wear protective gloves.**
>
> **Clinitest tablets (Part IIIB) contain sodium hydroxide. Avoid contact with skin, eyes, mucous membranes, and clothing. If possible, wear protective gloves.**

Procedure

I. ABSORPTION OF CARBOHYDRATES

1. With a short piece of string, **tightly** tie one end of a 10-cm (4-inch) dialysis bag.
2. Half fill the bag with a 1% cooked starch solution; then add 20% glucose solution until full. **Tightly** tie the other end with a piece of string.
3. Using a stirring rod, suspend the dialysis bag in a 400-mL beaker of deionized water as shown in Figure 17-3.

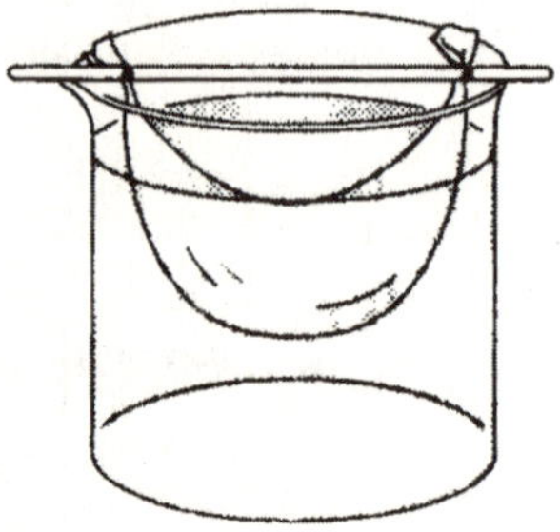

Figure 17-3 Dialysis bag suspended in a beaker of water.

4. Allow the bag to remain in the water for about 1-1/2 hours. Proceed to Part II.
5. Remove the bag, cut the string at one end, and pour the contents into a 150-mL beaker.
6. Test 1 mL of the solution inside the bag and 1 mL of the solution outside the bag with iodine solution as in Part IV. A positive test indicates the presence of starch.
7. Test 1 mL of each of the two solutions with Benedict's reagent as in Part IIIA. A positive test indicates the presence of glucose.

II. TEST FOR CARBOHYDRATES

Perform the tests of Parts II, III, and IV on each of the following solutions. Label four test tubes as follows:

No. 1: 1% glucose
No. 2: 1% fructose
No. 3: 1% sucrose
No. 4: 1% uncooked starch

Use a fresh solution for each test. Pour about 12 mL of each solution into test tubes numbered as above for use as stock reagents in Parts II, III, and IV.

A. Molisch Test

CAUTION: Handle concentrated sulfuric acid (Part IIA, Molisch test) with care. Concentrated sulfuric acid can burn and blister the skin. Wash you hands after use. This should be used in a hood. If possible, wear protective gloves.

1. Pour 5 mL of solution to be tested into a test tube and add 2 drops of Molisch reagent.

2. Carefully pour about 3 mL of concentrated sulfuric acid (**CAUTION**) into another test tube.

3. Use disposable laboratory gloves for this Step. Incline the first test tube and **slowly** pour the acid down the side of the tube so as to form two distinct layers. **Observe extreme caution in pouring the acid; it must be added slowly and allowed to run down the side of the tube! WASH YOUR HANDS IMMEDIATELY!**

4. Observe the color at the boundary of the two layers. A purple color indicates the presence of a carbohydrate. Dispose of the contents of the test tube as directed by your instructor.

B. Seliwanoff's Test

Seliwanoff's reagent (Part IIB) contains resorcinol, a highly irritating substance; it also contains hydrochloric acid. Wash your hands after use.

1. Pour 1 mL of the solution to be tested into a test tube.

2. **Use disposable laboratory gloves for this step.** Add 4 mL of Seliwanoff's reagent (**CAUTION**) and place the test tube in a boiling water bath on a hot plate for about 8 minutes. Wash your hands.

3. Remove the test tube and observe the color produced. Wash your hands. Dispose of the contents of the test tube as directed by your instructor.

III. TESTS FOR REDUCING SUGARS

A. Benedict's Test

1. To 1 mL of the solution to be tested add 5 mL of Benedict's solution.

2. Place the test tube in a boiling water bath on a hot plate for at least 5 minutes. Observe whether a color change is produced.

B. Clinitest Reagent

Clinitest tablets (Part IIIB) contain sodium hydroxide. Avoid contact with skin, eyes, mucous membranes, and clothing.

1. Place a test tube containing 1 mL of the solution to be tested in a rack and drop a Clinitest tablet into it. **CAUTION: do not handle the tablet with bare fingers.**

2. After 15 seconds, compare the color of the solution to the color chart.

3. Dispose of the contents of the test tube as directed by your instructor.

IV. IODINE TEST FOR POLYSACCHARIDES

1. Place 1 mL of the solution to be tested into a test tube.

2. Add a drop of 1% iodine solution; a blue color indicates the presence of starch.

3. Dispose of the contents of the test tube as directed by your instructor.

V. DIGESTION OF CARBOHYDRATES

A. Hydrolysis of Starch

Commercial α-amylase from human saliva (free from reducing sugars) is available and a solution of this solid can be prepared as the "commercial saliva" sample.

1. Crush a soda cracker in a test tube.

2. Add 10 mL of deionized water and place the tube in a warm-water bath, temperature 35° to 40°C, for 10 minutes.

3. Filter the suspension, collecting the filtrate in a second test tube, and divide it equally.

4. Test one part of for glucose as in Part IIIA and the second part for starch as in Part IV.

5. Test the sample of the commercial saliva solution for glucose as in Part IIIA.

6. With a mortar and pestle, grind up a cracker and add to about 15 mL of the "commercial saliva" solution in a small beaker. Place the beaker with the solution in a water bath on a hot plate and warm to about 35°C with continual mixing for about 7 minutes.

7. Prepare a funnel and filter paper for filtration. Use a test tube for your receiving vessel.

8. Pour the cracker mixture onto the filter paper, and wash with about 15 mL of deionized water, temperature 35° to 40°C. Collect about 3 mL of the filtrate.

9. Test 1 mL of the filtrate for glucose as in Part IIIA and 1 mL for starch as in Part IV.

10. Dispose of the soultions and residues as directed by your instructor.

B. Hydrolysis of Sucrose

1. Pour 10 mL of a 1% sucrose solution into a test tube.
2. Add 1 mL of 6 M HCl (**CAUTION!**) and place the test tube in a boiling-water bath over a hot plate for 10 minutes.
3. Pour approximately 1 mL of the solution into a second test tube and insert a strip of red litmus paper.
4. Neutralize the HCl by adding 10% sodium bicarbonate solution drop-by-drop until the red litmus just turns blue.
5. Test 1 mL of the solution with Benedict's reagent as in Part IIIA.

VI. THE ALPHA AND BETA FORMS OF D-GLUCOSE (Optional or Instructor's Demonstration)

1. Review the use of molecular models as described in Experiment 5. Use long-pegs for carbon-to-carbon bonds and carbon-to-oxygen single bonds. Carbon-to-hydrogen bonds are represented by short-pegs. Some model kits use two bent-pegs for double bonds; other kits use two spring-pegs for double bonds.
2. With your molecular model kit, assemble two models of the d-glucose molecule in its open-chain form. Do not insert the hydrogen atoms and the –OH groups on carbons 2, 3, 4, and 5 because this makes the structure too bulky and likely to fall apart. Connect all other atoms.
3. Without breaking the bonds between carbon atoms, arrange both structures as shown in Figure 17-4.

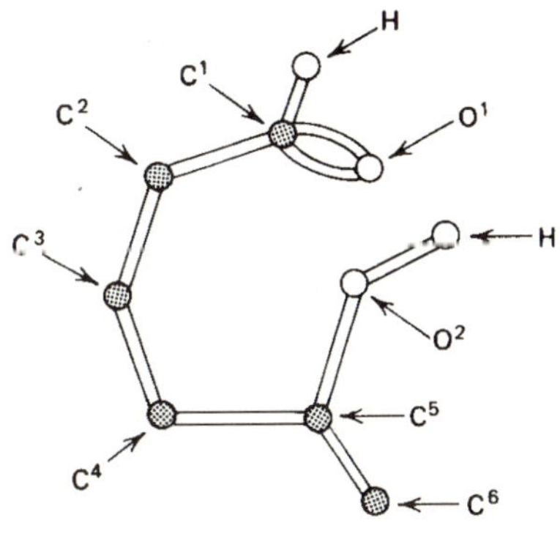

Figure 17-4 Model of D-glucose.

4. Lay the models flat on the laboratory bench. Of the two bonds between atoms C^1 and O^1, one will be resting on the bench and the second will be directly above it, as depicted in Figure 17-2. Place one model aside for later use and perform Steps 5 through 7 with the other model.
5. Detach the **upper** bent-peg (or spring-peg) from the oxygen (O^1) atom only. Do not detach the peg from C^1; atom C^1 should have a peg in every hole with nothing attached to the other end of one peg, and atom O^1 should have one vacant hole.

6. Detach the hydrogen and its peg from atom O^2 and attach it to O^1.

7. Rotate C^1 so that the unbonded peg can be bonded to O^2. Attach this bond.

8. Replace the spring-pegs (or bent-pegs) with long-pegs.

9. Repeat Steps 5 through 8 with the second model of glucose, but detach the lower spring-peg (or bent-peg) this time rather than the upper one.

10. Compare the two ring structures of glucose and sketch them on your Report Sheet.

Name ______________________ Section ____________ Date ____________

Prelab Questions for Experiment 17

______ 1. Which of the following compounds will give a positive Molisch test?
   a. maltose
   b. galactose
   c. cellulose
   d. all of the above (a, b, and c)
   e. none of the above (a, b, and c)

______ 2. Which of the following statements explains the basis for obtaining a positive Benedict's test?
   a. Secondary alcohols can be oxidized to form ketones.
   b. Primary alcohols can be oxidized to form aldehydes.
   c. Aldehydes are easily oxidized to form carboxylic acids.
   d. Tertiary alcohols cannot be oxidized easily.

______ 3. All carbohydrates are classified according to the
   a. compound produced upon hydrolysis.
   b. number of monosaccharides produced upon hydrolysis.
   c. number of carbon atoms they contain.
   d. functional group they contain.

______ 4. Seliwanoff's test is used to distinguish
   a. a ketose from an aldose.
   b. a monosaccharide from a disaccharide.
   c. a monosaccharide from a polysaccharide.
   d. a reducing sugar from a nonreducing sugar.

______ 5. When a dilute iodine solution is placed on a slice of bread, the solution turns dark blue because bread contains
   a. glucose.
   b. sucrose.
   c. starch.
   d. protein.

______ 6. Sucrose is a nonreducing sugar because
   a. it makes a person gain weight.
   b. it does not react with Cu^{2+}.
   c. it is a monosaccharide.
   d. it is a disaccharide.

7. Name the monosaccharides formed in the hydrolysis of:

   a. starch ______________________

   b. sucrose ______________________

8. What part of the body does the dialysis bag in Part I represent? ______________

Name ______________________ Section ____________ Date ____________

Report for Experiment 17

I. ABSORPTION OF CARBOHYDRATES

1. Data. Use a (+) to indicate a positive test and a (–) to indicate a negative test:

	Iodine Solution	*Benedict's Test*
Solution inside bag		
Solution outside bag		

2. Explain why you obtained the above results.

____

____

____

II. TEST FOR CARBOHYDRATES

1. Data:

Carbohydrate Solution	*Molisch (+ or –)*	*Seliwanoff's (Red or Pink)*	*Benedict's (+ or –)*	*Clinitest (%)*	*Iodine (+ or –)*
1% glucose					
1% fructose					
1% sucrose					
1% uncooked starch					

2. Explain your results of the Seliwanoff's test with sucrose solution.

____

____

3. A student tested starch with Benedict's solution and obtained a weak positive test. Explain.

____

____

V. DIGESTION OF CARBOHYDRATES

A. Hydrolysis of Starch

1. Data. Use a (+) to indicate a positive test and a (–) to indicate a negative test:

	Iodine Solution	*Benedict's Test*
Crushed cracker		
Chewed cracker		

2. Explain the test results.

Crushed cracker ____________________

____

Chewed cracker ____________________

____

3. Write a word equation to illustrate the reaction that occurs when starch is completely hydrolyzed.

____

B. Hydrolysis of Sucrose

1. Data. Use a (+) to indicate a positive test and a (–) to indicate a negative test:

	Benedict's Test
Sucrose (results from Part II)	
Sucrose after being hydrolyzed in HCl	

2. Explain why hydrolyzed sucrose gave different results from unhydrolyzed sucrose.

____

____

3. Write a word equation to illustrate the reaction that occurs when sucrose is hydrolyzed.

____

Name ________________________ Section ____________ Date ______________

VI. THE ALPHA AND BETA FORMS OF D-GLUCOSE (Optional)

1. Sketches of the two structures.

2. What class of compound is formed when the open-chain structure of D-glucose converts to the cyclic form? (Hint: consult your text.)

____

3. What kind of isomers, conformational or geometric, are α-glucose and β-glucose?

____

Name ______________________ Section ____________ Date ____________

Related questions for Experiment 17

1. After chewing on a piece of toast for five minutes, a student noticed a sweet taste in her mouth. Explain her observations.

____

____

2. Why is D-glucose (dextrose), rather than sucrose, given intravenously to patients?

____

____

3. Glucose in the urine may be a symptom of what disease?

____

4. Explain why infant formulas often contain mixtures of dextrins and maltose rather than starch.

____

____

5. a. What polysaccharide consists of α-D-glucose units? ____________________

   b. What polysaccharide consists of β-D-glucose units? ____________________

6. List two advantages in using the Clinitest reagent in preference to Benedict's reagent.

____

____

7. Why is it necessary that you digest (hydrolyze) starchy food?

____

8. Why is toast easier to digest that bread?

____

9. Why might cooked carrots taste sweeter than uncooked carrots?

____

Experiment 18
Proteins

Reference: *General, Organic, and Biological Chemistry: An Integrated Approach, Chapter 13.*

Purpose: You will perform protein identification tests, detect amino acids in orange juice, and investigate acid-base properties of proteins. You will identify factors that cause protein denaturation and determine the isoelectric point of casein, a typical protein.

Introduction

AMINO ACIDS

Proteins, essential components of the living cell, are high molecular weight compounds which, upon hydrolysis, produce amino acids. In a protein the amino acids are linked together by a series of peptide bonds. A **peptide bond** is an amide linkage; it is the functional group which can be **hydrolyzed** (split apart by water).

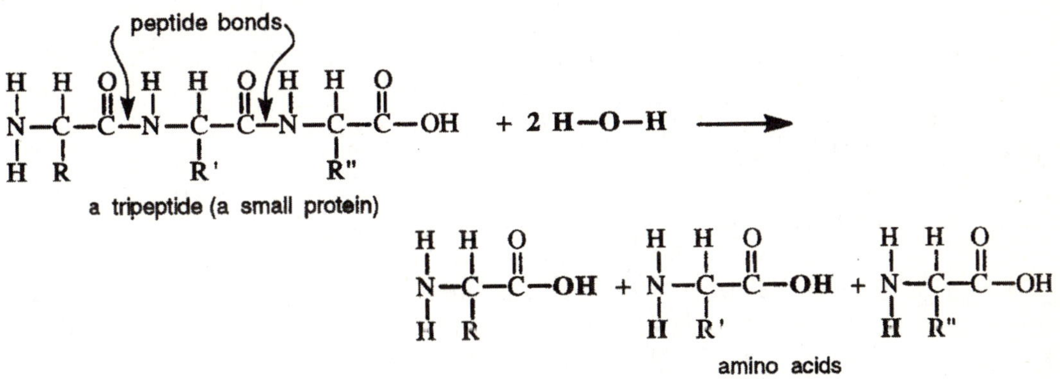

The group represented by R may be aliphatic, aromatic, or may contain another functional group. For example:

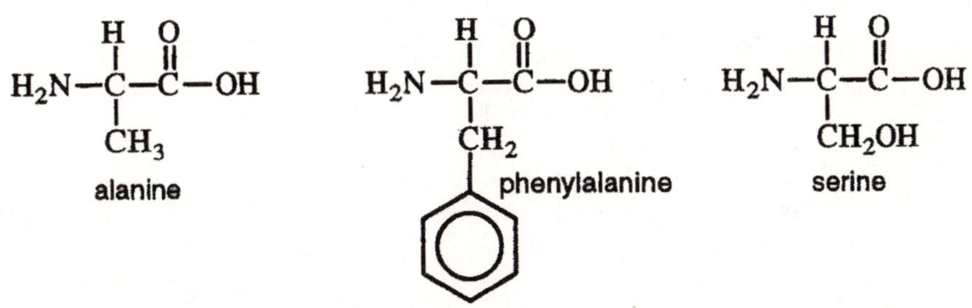

Human beings cannot synthesize ten of the twenty common amino acids found in protein; these **essential amino acids** must be supplied in the diet.

The presence of both the basic amino group ($–NH_2$) and the acidic carboxyl group (–COOH) accounts for the fact that amino acids (or proteins) in water solution can act as either acids or bases. Substances that can act as either acids or bases are **amphoteric**. In this experiment you will use indicators (phenolphthalein and methyl orange) to illustrate the amphoteric properties of casein, the principle protein found in cow's milk. Methyl orange appears red in a solution having a pH less than 3, and yellow in a solution of pH greater than 4. The indicator phenolphthalein is also used as a laxative in Ex-Lax®; it is colorless in a solution of pH less than 7.8 and pink in a solution having a pH greater than 9.6.

THE ISOELECTRIC POINT OF A PROTEIN

Amino acids (and proteins) contain basic amino groups; when an amino group accepts a proton in an acid-base reaction, the compound becomes positively charged. When the carboxylic acid group of an amino acid (or protein) donates a proton, the substance becomes negatively charged. At a specific pH, the compound will be neutral; this pH is the **isoelectric point** of the protein (or amino acid). Each protein has a characteristic isoelectric point. At this pH, the molecules coagulate and are removed from solution.

PROTEIN TESTS

There are several color tests that are used to detect proteins. These tests depend on the presence of a particular structure within the protein. A positive **biuret test**, for example, is given by all proteins, and requires the presence of two peptide linkages. When a protein in basic solution is treated with copper(II) sulfate ($CuSO_4$), a violet color is formed.

A second test, the **ninhydrin test**, is used extensively in chromatography to detect the presence of amino acids. In practice, the completed chromatogram is allowed to dry, sprayed with the ninhydrin reagent, and then placed in an oven at 100°C for approximately 5 minutes to develop a violet color, caused by the presence of the amino group.

PROTEIN STRUCTURE

The **primary structure** of a protein refers to the exact order in which the amino acids are linked together by means of peptide bonds. The three-dimensional configuration of segments of the protein chain is called its **secondary structure**; common secondary structures are the α-helix and the β-pleated sheet. The secondary structure is formed when amino acids hydrogen bond to other amino acids farther along the polypeptide chain. The **tertiary structure** is the three-dimensional shape of the entire polypeptide chain. Globular proteins, for example, are very tightly folded into a compact spherical form. This folding results from interactions between the R side chains of amino acids, and may involve hydrogen bonding as well as disulfide bondings, salt bridges, and hydrophobic interactions. Only proteins containing more than one chain, such as hemoglobin, have a **quaternary structure**; the manner in which the several peptide chains fit together describes a protein's quaternary structure.

DENATURATION OF PROTEINS

If the secondary, tertiary, or quaternary structures of a protein are disrupted, the protein loses its biological activity and is said to be **denatured**. Denaturation may result in coagulation with the protein being precipitated from solution. The factors that may cause the denaturation of proteins are pH, heat, certain organic solvents, heavy metal ions, alkaloidal reagents, reducing agents, and non-ionizing radiation. In this experiment you will study the denaturation of albumin, the protein found in egg white.

CAUTION: Handle the concentrated acids, hydrochloric, nitric, and sulfuric, with care. Avoid skin contact. Wash your hands after use. Wear gloves if possible.

Ninhydrin may be irritating to skin and mucous membranes. Wear gloves if possible.

Most organic compounds, such as ethanol, are flammable.

Procedure

I. DETECTION OF AMINO ACIDS IN ORANGE JUICE

CAUTION: Ninhydrin may be irritating to skin and mucous membranes.

Most organic compounds, such as ethanol, are flammable.

1. Pour 2 mL of orange juice into a test tube.
2. Add 4 mL of ethanol to precipitate the protein. (**CAUTION: ethanol is flammable!**)
3. Filter the solution. Only a few drops of the filtrate are needed for use in Step 4.
4. Review thin layer chromatography as described in Experiment 2, Part II. Use a capillary tube to apply the filtrate to a 2 × 7 cm silica TLC sheet. Develop the chromatogram in a 150-mL beaker. Your instructor will provide the developing eluent: a mixture of ethanol, ammonium hydroxide, and water. **Do not touch the TLC sheet with your fingers—use forceps.** Proceed to Part II while waiting for the chromatogram to develop.
5. After the eluent front has neared the top of the silica sheet, remove the chromatogram, mark the solvent front, and allow it to dry.
6. Dip the chromatogram in ninhydrin solution (or use ninhydrin spray) and place it in an oven heated to 100°C for about 5 minutes to develop the spots.
7. Circle the spots that appear and calculate the R_f values. Dispose of the eluent as directed by your instructor.

II. ACID-BASE PROPERTIES OF PROTEINS

1. Place 5 mL of deionized water in a test tube. Add a drop of 1% sodium hydroxide solution and 2 drops of phenolphthalein solution. Note the color.
2. Add a very small quantity of casein; then stopper and shake the test tube until there is a color change. Explain you results.
3. Pour 5 mL of deionized water into each of two test tubes; then add a drop of 1% hydrochloric acid and 5 drops of methyl orange solution to each. Note the color.
4. To one of the tubes add a small quantity of casein; then stopper and shake the tube periodically for about 10 minutes until there is a color change. Compare the color produced to the color of the other test tube from Step 3.

III. THE ISOELECTRIC POINT

1. Pour about 50 mL of skim milk or an aqueous solution of powdered milk into a 150-mL beaker. Record the pH of the milk. For best results use a pH meter; indicator paper having approximately a 4.5 to 8.5 range will provide satisfactory readings.

2. Place a stirring bar in the beaker and place the beaker on a magnetic stirrer. Adjust the stirrer so that the bar is rotating at moderate speed. Use a stirring rod if a magnetic stirrer is not available.

3. Add 0.1 M acetic acid drop-by-drop from a medicine dropper until the proteins (casein) begin to coagulate. The coagulation point can be detected by swirling the beaker and observing whether particles adhere to its side. Record the pH of the milk—this is the isoelectric point.

IV. DENATURATION OF PROTEINS

CAUTION: Handle the concentrated acids, hydrochloric, nitric, and sulfuric, with care. Avoid skin contact. Wash your hands after use. Wear gloves if possible.

Ninhydrin may be irritating to skin and mucous membranes. Wear gloves if possible.

Dispose of the contents of all test tubes as directed by your instructor.

A. pH

1. Pour about 3 mL of 2% albumin solution into each of three test tubes in a test-tube rack, number the tubes and add the following reagents: (**CAUTION!**) Wash you hands.
   No. 1: 1 mL of concentrated HCl
   No. 2: 1 mL of concentrated HNO_3
   No. 3: 1 mL of concentrated H_2SO_4

2. Stopper and shake each test tube. Record your observations. Wash your hands.

B. Heat

1. Pour about 5 mL of albumin solution into a test tube and heat the tube in a boiling-water bath over a hot plate for about 10 minutes.

2. Repeat the test with a 5-mL sample of skim milk or powdered milk solution. Carefully note whether any coagulants have appeared.

C. Organic Solvents

Most organic compounds, such as ethanol, are flammable.

1. Pour about 5 mL of albumin solution into a test tube and add 5 mL of 95% ethanol. **CAUTION: ethanol is flammable. No flames present!**

2. Stopper the test tube and shake. Compare its appearance with the untreated albumin solution.

D. Heavy Metal Ions

1. Pour 3 mL of albumin solution into each of two test tubes and place them in a rack. Number the test tubes, and add 10 drops of the following reagents:
   No. 1: 1% silver nitrate ($AgNO_3$) solution
   No. 2: 1% sodium chloride (NaCl) solution

2. Stopper and shake each test tube. Record your observations.

E. Alkaloidal Reagents

1. Pour about 5 mL of albumin solution into a test tube and add about 1 mL of a 10% tannic acid solution.

2. Stopper and shake each test tube. Record your observations.

V. TEST FOR PROTEINS

A. Biuret Test

1. Pour 5 mL of albumin solution into a test tube and add an equal volume of 10% NaOH solution.

2. Stopper and shake the test tube. Add about 5 drops of 0.5% $CuSO_4$ solution. Record your observations.

3. Dispose of the contents of the test tube as directed by your instructor.

B. Ninhydrin Test

CAUTION: Ninhydrin may be irritating to skin and mucous membranes.

1. Obtain a circle of filter paper from your instructor. **Handle the paper with forceps.**

2. Moisten your thumb with water, then press it firmly on 3 or 4 places on the filter paper.

3. Expose the spots to ninhydrin solution using a spray or by placing the ninhydrin solution on the spots with a medicine dropper. (**CAUTION!**)

4. Place the paper in an oven heated to 100°C or heat the paper over a hot plate for about 5 minutes. Record your observations. Attach the filter paper to your Report Sheet.

Name ______________________ Section ____________ Date ____________

Prelab Questions for Experiment 18

True or False:

________ 1. All amino acids will produce a positive biuret test.

________ 2. All amino acids will produce a positive ninhydrin test.

Multiple choice:

________ 3. At the isoelectric point of a protein,
   a. the pH of the solution will be 7.
   b. the number of positive charges on the protein will equal the number of negative charges.
   c. the pH of the solution will always be greater than 7.
   d. the pH of the solution will always be less than 7.

________ 4. Ethanol is added to the orange juice in Part I of this experiment to
   a. precipitate the protein.
   b. detect the amino acids.
   c. develop the chromatogram.
   d. emulsify the orange juice.

5. A chromatogram is performed on a mixture of amino acids. Given the following data, calculate the R_f value for each amino acid.

Eluent distance	4.5 cm	
		R_f Value
Amino acid "A" distance	0.5 cm	________
Amino acid "B" distance	1.2 cm	________
Amino acid "C" distance	3.6 cm	________

Calculations:

Name ______________________________ Section ____________ Date ______________

Report for Experiment 18

I. DETECTION OF AMINO ACIDS IN ORANGE JUICE

1. Attach your chromatogram here.

2. Label each amino acid you have circled (A, B, C, etc.) and calculate the R_f value for each.

		R_f *Value*
Eluent distance	____________	
Amino acid "A" distance	____________	____________
Amino acid "B" distance	____________	____________
Amino acid "C" distance	____________	____________
Amino acid "D" distance	____________	____________

Calculations:

II. ACID-BASE PROPERTIES OF PROTEINS

1. Data:

   a. Color of phenolphthalein/NaOH solution

   —before adding casein ____________________

   —after adding casein ____________________

   b. Color of methyl-orange/HCl solution

   —before adding casein ____________________

   —after adding casein ____________________

2. Why did the solutions change color?

____

3. Which property of proteins is illustrated by the above?

____

4. Why was methyl orange used in place of phenolphthalein in the second part of this procedure?

____

III. THE ISOELECTRIC POINT OF CASEIN

1. Data:

   Initial pH of milk ____________

   Final pH of milk ____________

2. What is the isoelectric point of casein? ____________

IV. DENATURATION OF PROTEINS

A. pH

1. Observations

   a. Tube No. 1: concentrated HCl

   ____

   ____

   b. Tube No. 2: concentrated HNO_3

   ____

   ____

   c. Tube No. 3: concentrated H_2SO_4

   ____

   ____

2. Which structure or structures (primary, secondary, tertiary, or quaternary) of the protein were disrupted in this procedure? Refer to your text.

____

Name ______________________ Section ____________ Date ____________

B. Heat

1. Observations of heated solutions:

   a. Albumin solution

   ____

   ____

   b. Skim milk

   ____

   ____

2. Which structure or structures of the protein were disrupted in this procedure?

   ____

C. Organic Solvents

1. Observations of ethanol added to albumin

   ____

2. Which structure or structures of the protein were disrupted in this procedure?

   ____

D. Heavy Metal Ions

1. Observations:

   a. Tube No. 1: $AgNO_3$ solution

   ____

   b. Tube No. 2: NaCl solution

   ____

2. Explain the differences in your results.

   ____

   ____

3. Which structure or structures of proteins were disrupted in this procedure?

   ____

E. Alkaloidal Reagents

1. Observations of tannic acid added to albumin.

2. Which structure or structures of the protein were disrupted in this procedure?

V. TEST FOR PROTEINS

A. Biuret Test

Observations

B. Ninhydrin Test

1. Observations

2. Attach your filter paper here.

Name ______________________ Section ____________ Date ____________

Related Questions for Experiment 18

1. Explain why precipitation takes place at the isoelectric point of an amino acid or protein.

____

____

2. Why are lead compounds poisonous?

____

____

____

3. a. Explain why dilute silver nitrate is applied to the eyes of newborn infants.

____

____

   b. Explain why dilute, rather than concentrated, silver nitrate must be used when applied to the eyes.

____

____

4. Why is isopropyl alcohol a good disinfectant?

____

____

5. Why are tannic acid and picric acid used in commercial products for the treatment of burns?

____

____

6. Based on your laboratory observations, how might ninhydrin be used in forensic chemistry to solve crimes?

____

____

____

7. a. NutraSweet®, the synthetic sweetener aspartane, is a dipeptide (protein) that is used in low-calorie beverages. Explain why beverages containing NutraSweet® lose their sweetness after several months of storage.

____

____

b. Explain why NutraSweet® loses its sweetness when heated and cannot be used in cooking.

____

____

Experiment 19
Enzymes

Reference: *General, Organic, and Biological Chemistry: An Integrated Approach, Chapter 13.*

Purpose: You will perform tests to determine the chemical nature of enzymes and the specificity of enzyme action. You will investigate the effects of temperature, pH, substrate concentration, and enzyme concentration on enzyme activity.

Introduction

Enzymes are water-soluble proteins that function as biological catalyst for the reactions involved in metabolism. They may be **simple proteins** composed entirely of amino acids or may be **conjugated proteins** that require a nonprotein group for their biological activity. Enzymes are similar to inorganic catalysts in that they catalyze reactions by lowering the activation energy. However, they differ from inorganic catalysts in their specificity. For example, platinum will catalyze several different types of reactions, but a given enzyme will catalyze only one type of reaction. This **specificity** of enzymes is explained by the **lock-and-key theory**. This theory assumes that there is a specific area on the surface of the enzyme, called the **active site**, to which the substance acted upon (the **substrate**) attaches during a reaction. The configuration (shape) of the enzyme's active site and substrate must be complementary in much the same way as a key fits into a lock. The configuration of the lock is specific for only one key; no other keys will turn the lock. In summary, enzymes are proteins whose catalytic activity is dependent upon their specific three-dimensional shape.

Factors such as pH, temperature, solvents, and salt concentrations can alter the secondary, tertiary, and quaternary structure of a protein. Such factors, therefore, will have an effect on the activity levels of enzymes. The **optimum pH** of an enzyme is that pH at which it will function most effectively, and the optimum temperature is the temperature of most efficient catalytic activity.

The three enzymes used in this experiment—urease, pepsin, and amylase—are representative of all enzymes; they were selected because of their availability and the ease with which their activity may be observed in simple laboratory experiments.

Urease catalyzes a very specific chemical reaction: the hydrolysis of urea.

$$H_2N-\overset{\overset{\displaystyle O}{\|}}{C}-NH_2 + H_2O \xrightarrow{\text{urease}} 2\,NH_3\,(g) + CO_2\,(g)$$

One of the products of this reaction, ammonia, increases the pH of the urease solution enough so that the reaction may be followed with indicators such as phenol red, which is yellow in acid solution and red in base solution.

Pepsin is found in the stomach and urine; it hydrolyzes peptide bonds between two aromatic amino acids or between glutamic acid and an aromatic amino acid. Its optimum pH varies from 2.0 to 4.5.

Amylase, found in the urine and saliva, hydrolyzes starch (a polysaccharide) to form maltose (a disaccharide). The digestion (hydrolysis) of starch may be followed by periodically testing the solution with an iodine solution, which produces an identifying color as illustrated in the following word equation.

$$\underset{\text{(blue)}}{\text{starch}} \xrightarrow{\text{amylase}} \underset{\text{(blue)}}{\text{amylodextrin}} \longrightarrow \underset{\text{(red)}}{\text{erythrodextrin}} \longrightarrow \underset{\text{(colorless)}}{\text{achrodextrin}} \longrightarrow \underset{\text{(colorless)}}{\text{maltose}}$$

Because the activity of amylase is increased in the presence of chloride ion, the urine solution (our source of amylase in the experiment) is mixed with a sodium chloride solution before studying its properties. Amylase is also activated by calcium ion found in the urine.

Procedure

I. THE CHEMICAL NATURE OF ENZYMES (Biuret Test)

1. Pour about 3 mL of 1% urease solution into a test tube and add an equal volume of 10% NaOH.

2. Stopper and shake the test tube. Add about 5 drops of 0.5% $CuSO_4$ solution. Record your observations. Refer to Experiment 18 for discussion on the Biuret test.

II. SPECIFICITY OF ENZYME ACTION

1. Label three test tubes, and add the following reagents.
   - No. 1: 8 mL of deionized water + 3 drops of phenol red indicator (control)
   - No. 2: 3 mL of 1% **urease** solution + 5 mL of 2% urea solution + 3 drops of phenol red indicator
   - No. 3: 3 mL of 1% **pepsin** solution + 5 mL of 2% urea solution + 3 drops of phenol red indicator

2. Stopper and shake each test tube. Observe the color in each test tube.

III. THE EFFECT OF SUBSTRATE CONCENTRATION

1. Label three small Erlenmeyer flasks and add the following reagents.
   - No. 1: 30 mL of deionized water (control)
   - No. 2: 10 mL of 2% urea solution + 20 mL of deionized water + 0.1 g urease
   - No. 3: 5 mL of 2% urea solution + 25 mL of deionized water + 0.1 g urease

2. Place the flasks in a water bath or an oven maintained at 40–45°C.

3. After 15 minutes remove the flasks, empty the contents of each into three separate numbered 100-mL beakers, and add 3 drops of universal indicator solution. Observe the color in each of the flasks and record the pH. Refer to Experiment 11 for the discussion on universal indicator.

4. If available, use a pH meter to measure the pH of each solution.

IV. TEMPERATURE AND ENZYME ACTIVITY

We require that disposable latex or vinyl gloves be worn while handling the urine solution.

Prepare a urine solution for use in Parts IV, V, and VI by mixing 30 mL of commercial urine with 30 mL of 0.9% NaCl.

1. Pour 20 mL of freshly prepared 0.075% starch solution into each of three numbered 50-mL Erlenmeyer flasks.

2. Place flask No. 1 in a water bath, maintained at 55–60°C, using a hot plate. Place flask No. 2 in a water bath, maintained at 40°C, using a hot plate. Place flask No. 3 in an ice-water bath.

3. Allow 5 minutes for each of the solutions to reach the temperature of the water bath.

4. Add 15 mL of urine solution to each of the three flasks. **Mix well by swirling the flasks.** Note the time at which the urine was added.

5. After 5 minutes test for the presence of starch in each of the solutions as follows. Place three drops of freshly prepared 0.025% iodine solution in a small (10- × 75-mm) test tube. Using a medicine dropper, add 3 drops of urine-starch solution. Use separate test tubes and a different medicine dropper for each sample. Compare the colors of the three solutions.

6. After testing the samples, return each Erlenmeyer flask to its respective water bath. Repeat the tests at 5 minute intervals until one of the solutions produces the red color characteristic of erythrodextrin. Record the time for a change to take place.

7. Dispose of the solutions as directed by your instructor.

V. PH AND ENZYME ACTIVITY

We require that disposable latex or vinyl gloves be worn while handling the urine solution.

1. Label three test tubes in a rack, number them, and add the following reagents:
   No. 1: 6 mL of 0.075% starch solution + 3 mL of buffer (pH 6)
   No. 2: 6 mL of 0.075% starch solution + 3 mL of buffer (pH 7)
   No. 3: 6 mL of 0.075% starch solution + 3 mL of buffer (pH 8)

2. Add 2 mL of urine solution and place the tubes in a hot-water bath or oven maintained at approximately 40°C.

3. After 5 minutes test for the presence of starch as described in Part IV, Step 5.

4. Continue the tests at 5-minute intervals. Record the time required for one of the three solutions to produce the red color characteristic of erythrodextrin.

5. Continue the tests until one of the solutions turns the iodine solution colorless, indicating the presence of achrodextrin.

6. Dispose of the solutions as directed by your instructor.

VI. THE EFFECT OF ENZYME CONCENTRATION

We require that disposable latex or vinyl gloves be worn while handling the urine solution.

1. Dilute the urine stock solution of Part IV by adding 9 mL of deionized water to 1 mL of stock solution.

2. Place two test tubes in a rack, number them, and add the following reagents:
   No. 1: 6 mL of 0.075% starch solution + 3 mL of buffer (pH 7) + 2 mL of **diluted** urine solution prepared in Step 1
   No. 2: 6 mL of 0.075% starch solution + 3 mL of buffer (pH 7) + 2 mL of urine solution

3. Place the two test tubes in a water bath or oven maintained at approximately 40°C. Note the time.

4. After 10 minutes test for the presence of starch in each of the two test tubes as described in Part IV, Step 5.

5. Continue the tests at 5-minute intervals until one of the two solutions produces the erythrodextrin red color.

6. Dispose of the solutions as directed by your instructor.

Name ______________________ Section ____________ Date ______________

Prelab Questions for Experiment 19

1. Starch is being hydrolyzed in the presence of amylase to form erythrodextrin. How would you test for the presence of erythrodextrin?

____

____

2. When urea is hydrolyzed, the pH of the solution rises. Why?

____

____

3. In Part III, The Effect of Substrate Concentration, how can you determine whether solution 2 or 3 reacts faster?

____

____

4. What other body fluid could have been substituted for the urine solution in Parts IV through VI of this experiment and still have yielded the same results?

____

____

5. Why is the Biuret test performed on the urease solution in Part I?

____

Name ______________________ Section ____________ Date ______________

Report for Experiment 19

I. THE CHEMICAL NATURE OF ENZYMES (Biuret Test)

1. Observations

____

____

____

2. Based on your observations, to what class of compounds do enzymes belong?

____

II. SPECIFICITY OF ENZYME ACTION

1. Observations:

   a. No. 1—water (control)

   ____

   b. No. 2—urease + urea

   ____

   c. No. 3—pepsin + urea

   ____

2. a. Which of the two enzymes catalyzed a chemical reaction?

   ____

   b. How do you know a reaction took place?

   ____

   c. What were the products of the reaction?

   ____

3. What was the purpose of the control test tube (No. 1—water)?

____

III. THE EFFECT OF SUBSTRATE CONCENTRATION

1. Observations:

   a. No. 1—water (control) pH ________________

   ____

   b. No. 2—10 mL of 2% urea + urease pH ________________

   ____

   c. No. 3—5 mL of 2% urea + urease pH ________________

   ____

2. a. Which of the above reactions proceeded at the fastest rate?

   ____

   b. Explain, with the use of an equation, how the pH enabled you to arrive at your conclusion.

   ____

   ____

   ____

IV. TEMPERATURE AND ENZYME ACTIVITY

1. Data:

Solution	*Temperature*	*Time urine was added*	*Time iodine solution is red*	*Time iodine solution is colorless*
No. 1				
No. 2				
No. 3				

2. Which reaction temperature hydrolyzed starch the fastest? ________________

3. Explain in terms of molecular structure why the other two solutions reacted more slowly.

   ____

   ____

Name ______________________ Section ____________ Date ____________

V. PH AND ENZYME CONCENTRATION

1. Data:

Solution	*pH*	*Time urine was added*	*Time iodine solution is red*	*Time iodine solution is colorless*
No. 1	6			
No. 2	7			
No. 3	8			

2. In which of the three solutions was the starch hydrolyzed most rapidly? ____________

3. Explain in terms of molecular structure why the other two solutions reacted more slowly.

____

____

VI. THE EFFECT OF ENZYME CONCENTRATION

1. Data:

Solution	*Time urine was added*	*Time iodine solution is red*	*Time iodine solution is colorless*
Diluted urine solution			
Urine solution			

2. Which solution hydrolyzed starch the fastest?

____

3. Explain your results.

____

____

Name ______________________________ Section ____________ Date ____________

Related Questions for Experiment 19

1. Explain how your experimental data in Parts II through VI support the lock-and-key theory of enzyme action.

____

____

____

____

____

____

____

____

2. Why will lead nitrate solution slow or stop the hydrolysis of starch?

____

____

3. Why will the addition of oxalate ion to the urine slow or stop the hydrolysis of starch? (Hint: see Experiment 16.)

____

____

____

Experiment 20

DNA and RNA*

Reference: *General, Organic, and Biological Chemistry: An Integrated Approach, Chapter 14.*

Purpose: With special sets of molecular models you will simulate the replication of DNA, the transcription of DNA to mRNA, and the translation of mRNA in protein synthesis.

Introduction

A detailed discussion of the replication of DNA, transcription DNA to mRNA, and the translation of mRNA is provided in you text. The following discussion is intended only to review some of the more important aspects of DNA and RNA functions as applied to your DNA kit.

THE REPLICATION OF DNA

Deoxyribonucleic acid (DNA) is the hereditary molecule in all cells. The double helix structure of DNA has a backbone consisting of a five-carbon sugar (deoxyribose) and phosphate groups connected by ester linkages. Attached to the backbone are four nitrogen bases. The sequence of the four bases is the genetic code, the hereditary information of the cell. The bases of one strand of the double helix are hydrogen bonded to the bases of the other strand in a complementary manner. The base thymine (T) is always hydrogen bonded to adenine (A); cytosine (C) is always hydrogen bonded to guanine (G).

A **nucleoside** is the combination of a sugar with a nitrogen base. A sugar, a nitrogen base, and a phosphate group combine to form a **nucleotide**. DNA and RNA are polymers of nucleotides in which the sugars are deoxyribose and ribose, respectively.

When a cell divides, each daughter cell receives DNA identical to the parent cell through the process of **replication**. During replication of DNA, the double helical structure of DNA unwinds and each strand serves as the pattern for the synthesis of a new complementary strand of DNA.

$$\underset{\text{(in parent cell)}}{\text{DNA}} \xrightarrow{\text{Replication}} \underset{\text{(in two daughter cells)}}{\text{DNA} + \text{DNA}}$$

The DNA model kit used in this experiment consists of interlocking brightly colored plastic components, representing the following structures: phosphates, five-carbon sugars, nitrogen bases,

*The DNA kits used in this experiment were designed by Ken Zietlow and Richard Brady and are distributed by Wards Biology, 5100 N. Henrietta Road; P.O.B. 92912; Rochester, NY 14692; Product No. 81W0140; www.wardsci.com. This experiment is based on a series of exercises in an accompanying *Student Guide for DNA Model Kit* and is published by permission of the authors.

amino acids, hydrogen, and hydroxyl groups. An itemized list of pieces is provided in Part I, Step 2 of the Procedure section. Ester linkages are represented by tightly interlocked pieces. Hydrogen bonding between bases is denoted by loose-fitting, but not interlocking, complementary structures.

In Part I of this experiment you will construct a model based upon a specific base sequence in a DNA strand. You will then assemble a second complementary strand of the DNA structure, linking together the bases that hydrogen bond to one another. The two strands will then be separated, and each of the two daughter strands will be used to assemble new DNA structures.

TRANSCRIPTION OF DNA TO RNA

Each cell must synthesize protein and each protein must have a specific order of amino acids (the primary structure of the protein). The sequence of nitrogen bases in DNA contains the genetic information required to assemble the amino acids in the correct order. Another nucleic acid, **ribonucleic acid (RNA)**, copies the genetic information from DNA and uses that information to direct the synthesis of protein in the cell's ribosomes. Like DNA, RNA has a backbone made of a five-carbon sugar with ester linkages to phosphate groups; however, the sugar in RNA is ribose rather than deoxyribose. Four nitrogen bases are attached to the backbone; three of the bases (adenine, cytosine, and guanine) are the same as in DNA. The RNA base uracil replaces the DNA base thymine. Several types of RNA exist, usually with a single-strand of nucleic acid.

The preliminary step in protein synthesis is a **transcription** of the genetic information from a segment of DNA to **messenger RNA (mRNA)**. This step is similar to the replication of DNA but only a segment of the DNA chain is copied, not the entire chain.

In Part II of this experiment, you will use models to demonstrate transcription of DNA to mRNA. During transcription the nitrogen bases of DNA hydrogen bond to the nitrogen bases of mRNA. Thymine (T) is always hydrogen bonded to uracil (U) and cytosine (C) hydrogen bonds only to guanine (G).

THE TRANSLATION OF MRNA (PROTEIN SYNTHESIS)

In the preliminary step of protein synthesis, genetic information was transcripted from DNA to mRNA in the nucleus of the cell. After being synthesized, the mRNA migrates to the cell's ribosomes.

Another form of RNA, **transfer RNA (tRNA)**, is much smaller than other nucleic acids. Each tRNA molecule is specifically designed for a particular amino acid. The amino acid becomes attached to the end of its tRNA molecule. One after another, tRNA molecules hydrogen bond to mRNA; the amino acids carried by the tRNA are assembled in sequence to make a protein. The order of the amino acid assembly is determined by the sequence of nitrogen bases in the mRNA. A sequence of three nitrogen bases on mRNA is a **codon**. The nitrogen bases of the mRNA codon hydrogen bond to complementary nitrogen bases in the **anticodon** of the tRNA. Most of the 20 common amino acids have more than one codon. For example, UUU and UUC are both codons for the amino acid phenylalanine.

To summarize, the preliminary step of protein synthesis, **transcription**, copies genetic information from DNA to RNA. The actual synthesis of protein occurs during **translation** when the sequence of nitrogen bases is translated into the sequence of amino acids of the protein. The key to the translation is the matching of specific amino acids to the codons of the mRNA using complementary anticodons on tRNA.

$$\text{DNA} \xrightarrow{\text{Transcription}} \text{RNA} \xrightarrow{\text{Translation}} \text{Protein}$$

Part III of this experiment illustrates the repeated steps that take place between tRNA and mRNA in synthesizing a polypeptide (protein).

Procedure

I. THE REPLICATION OF DNA

1. Obtain a DNA kit from your instructor. Work in pairs because the components of two kits will be needed for this experiment.

2. Separate the kit into its components by letter. You should have the following parts in each kit:

Symbol	*Quantity*	*Part Name*
A	24	Deoxyribose
B	12	Ribose
C	24	Phosphate
D	6	Uracil (U)
E	6	Thymine (T)
F	6	Guanine (G)
G	6	Cytosine (C)
H	2	Hydrogen
J	6	Adenine (A)
OH	2	Hydroxyl group
Amino acids	2	Two different amino acids of different color

3. Remove one each of Parts D, E, F, G, and J (uracil, thymine, guanine, cytosine, and adenine).

4. Test to see which **pairs** of the above parts hydrogen bond by attempting to join the pieces together as in a jig-saw puzzle. Try every combination possible.

5. With the use of parts from two kits, use A, C, E, F, G, and J to assemble 48 separate DNA nucleotides. Remember that each nucleotide consists of a deoxyribose sugar, a phosphate, and a nitrogen base. See Figure 20-1.

Quantity	*DNA Nucleotides*
12	phosphate + thymine + deoxyribose
12	phosphate + adenine + deoxyribose
12	phosphate + guanine + deoxyribose
12	phosphate + cytosine + deoxyribose

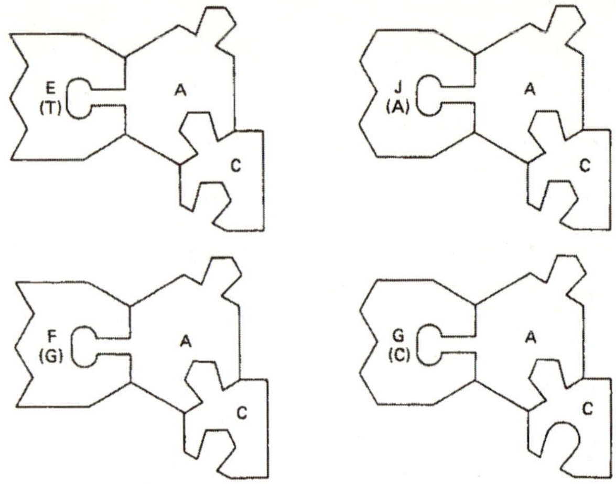

Figure 20-1 DNA molecules.

6. Arrange all nucleotides so that the phosphate groups (C) are on the right side of deoxyribose (A).

7. Attach 12 of these nucleotides end to end so that the sequence of bases matches the diagram of Figure 20-2. The letter symbol is that of the bases, not the code symbol for each part. See Step 2.

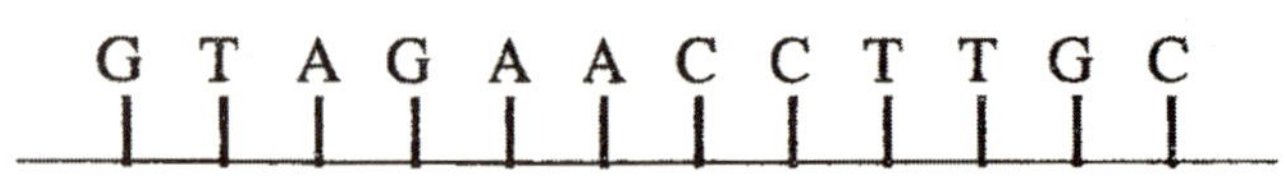

Figure 20-2 Base sequence in a DNA strand.

8. The 12 remaining nucleotides will now be used to construct the second strand of the DNA molecule. Begin linking together the bases that hydrogen bond and then attach the deoxyribose and phosphate structures of all nucleotides. You should now have a ladderlike arrangement, which represents the double-stranded DNA.

9. Separate the two strands of the DNA molecule, keeping each strand intact.

10. Using the 24 remaining nucleotides, construct two DNA molecules as you did earlier in Step 8. You have simulated the replication of DNA.

II. TRANSCRIPTION OF DNA TO MRNA

1. Disassemble one of the DNA molecules that you constructed in Part I. Completely separate the individual pieces (sugars, phosphates, and bases), and group similar molecules together.

2. Assemble 24 RNA nucleotides as indicated below. Make sure that the phosphate groups are on the right side of the ribose. See Figure 20-3. These nucleotides will constitute the RNA molecule pool in the cell.

Quantity	*RNA Nucleotides*
6	phosphate + uracil + ribose
6	phosphate + guanine + ribose
6	phosphate + adenine + ribose
6	phosphate + cytosine + ribose

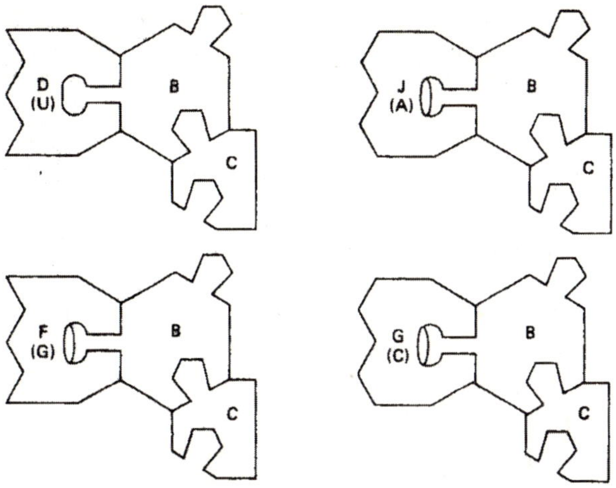

Figure 20-3 RNA nucleotides.

3. Construct a tRNA molecule by linking three nucleotides together as illustrated in Figure 20-4. The nucleotide should contain the bases GUA.

4. Construct three other tRNA molecules containing the bases GAA, CCU, and UGC.

5. Attach the amino acids, with H and OH on each end, to each of the four sets of tRNA molecules corresponding to the anticodons listed below. See Figure 20-4. Place the molecules aside for later use.

Anticodon	*Amino acid*
GUA	histidine
GAA	leucine
CCU	glycine
UGC	threonine

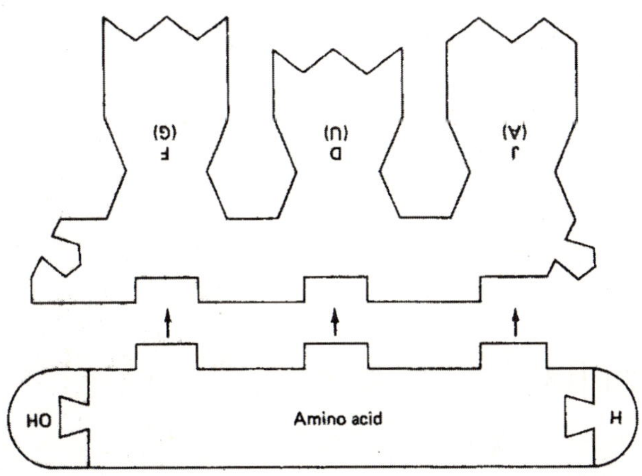

Figure 20-4 A tRNA molecule, showing how amino acid attaches.

6. Separate the double-stranded DNA as before. With the remaining nucleotides that you have constructed in Step 2, match the bases of the RNA nucleotides with the original strand of DNA. Remember, "matching bases" form hydrogen bonds.

7. Connect the RNA nucleotides. You have now constructed an mRNA molecule.

8. Place the mRNA molecule to one side and attach the two strands of DNA as before. The DNA molecule is back in its original state, capable of taking part in another transcription process if needed.

III. THE TRANSLATION OF MRNA (PROTEIN SYNTHESIS)

1. Position the mRNA molecule you have constructed in Part II on your laboratory table so that you can easily "match bases" as illustrated in Figure 20-5.

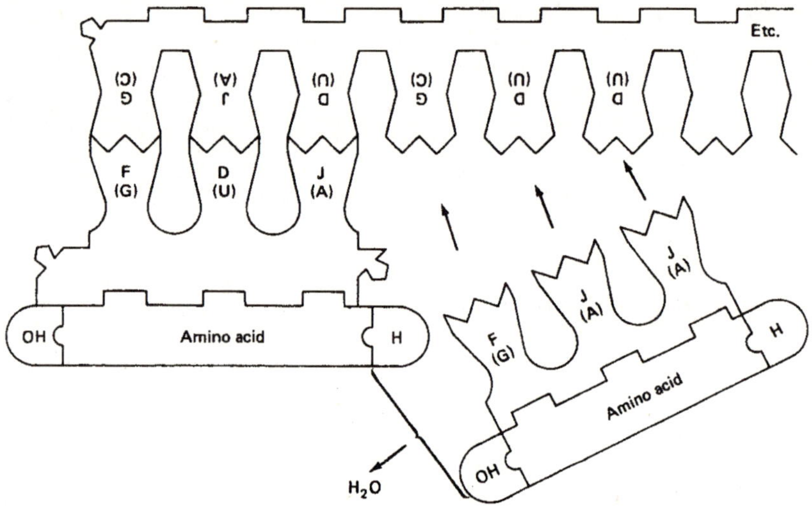

Figure 20-5 Translation of mRNA to polypeptide. The codons are written "upside down" on the mRNA molecule because the codon/anticodon pairings occur in an antiparallel direction.

2. Match anticodons on amino acid-carrying tRNAs (from Part II, Step 5) with the codons on the mRNA. You should have four amino acids placed side by side.

3. Construct a polypeptide by "splitting off" water molecules (H and OH) and attaching the amino acid fragments. Remove the polypeptide molecule. As each amino acid is removed from a tRNA and attached to the growing polypeptide, the tRNA is free to return to the amino acid pool to pick up another amino acid. This process replicates the synthesis of a polypeptide using the DNA genetic code. The sequence of amino acids in the polypeptide was determined by the sequence of nucleotides in DNA.

Name ______________________ Section ____________ Date ____________

Prelab Questions for Experiment 20

1. List the nitrogen bases found in DNA.

2. List the nitrogen bases found in RNA.

3. How is an ester linkage identified in the models used in this experiment?

4. How is a hydrogen bond identified in the models used in this experiment?

5. Which part of the experiment illustrates the synthesis of a protein?

Name ______________________________ Section ____________ Date ______________

Report for Experiment 20

I. THE REPLICATION OF DNA

1. List the nitrogen base pairs that hydrogen bond.

____

____

____

____

2. List the base sequence in the second DNA strand.

3. a. Does your DNA model represent the true geometry of the DNA molecule? ________

   b. If not, what is the major difference in structure?

____

____

II. TRANSCRIPTION OF DNA TO MRNA

1. List the base sequences on the mRNA molecule of Step 7.

2. What are the three-base sequences on the mRNA molecule called?

____

III. THE TRANSLATION OF MRNA (PROTEIN SYNTHESIS)

1. Using the standard method of illustrating polypeptide structure (e.g., Ala-Glu-Ser-Val), write the polypeptide formed in this experiment.

____

2. What are the three-base sequences on the tRNA called?

____

3. What happens to the mRNA after the polypeptide is synthesized?

____

____

Name ______________________ Section ____________ Date ______________

Related Questions for Experiment 20

1. Suppose that the base sequence for one strand of DNA in Part I of this experiment was as follows:

G A A G T A T G C C C T

a. List the base sequence in the second DNA strand.

b. List the base sequence in the mRNA strand that is formed from the DNA strand shown above.

c. List the codons in the mRNA strand of Question 1b.

____________ ____________ ____________ ____________

d. List the anticodons in the four tRNA molecules that would hydrogen bond to the mRNA of Question 1b.

____________ ____________ ____________ ____________

e. Using the standard method of illustrating polypeptide structure, write the polypeptide that would be formed on the mRNA strand from Question 1b.

____

2. a. What is a mutation?

____

____

b. Give at least two examples of how a mutation can occur.

____

____

____

____

3. a. Must two mRNA molecules be identical to produce the same polypeptide?

____

b. Explain.

____

____

____

4. With the help of reference sources, list some of the diseases associated with mutations, and briefly describe their symptoms.

____

____

____

____

____

____

____

____

____

Experiment 21
Urine

Purpose: You will investigate the properties of a commercial urine sample with respect to pH, specific gravity, and some inorganic and organic compounds. You will diagnose a simulated "pathological" urine specimen for the above properties.

Introduction

Urine is an aqueous solution of various organic and inorganic substances that are the waste products of metabolism or products from ingested foods and fluids. The adult human body excretes an average of 800 to 2000 milliliters of urine in 24 hours. More urine is formed during the waking hours than during an equivalent period of sleep. The total quantity of waste products excreted during the two periods is about the same, but the night urine is more concentrated. The average composition of urine of a healthy adult is as follows:

Water	95%	Sodium, Na^+	0.35%
Urea	2%	Potassium, K^+	0.15%
Uric acid	0.05%	Ammonium, NH_4^+	0.04%
Creatinine	0.075%	Calcium, Ca^{2+}	0.015%
Glucose	0%	Magnesium, Mg^{2+}	0.006%
Protein	0%	Chloride, Cl^-	0.60%
Ketone bodies	0%	Sulfate, SO_4^{2-}	0.18%
		Phosphate, HPO_4^{2-}, $H_2PO_4^{-1}$	

Urea is the major organic component in urine. Excretion of urea is the principal method for removing nitrogen from the body.

The clinical purpose for analyzing urine is to determine the existence of a body malfunction. The presence of unusually large or small quantities of normal metabolic end products or the existence of abnormal substances in the urine may indicate a metabolic disorder. The disturbance may be a malfunction directly associated with the kidney, or it may be a condition in which the kidneys function normally but excrete a specific metabolic product associated with a particular disease.

For example, **phenylketonuria**, or **PKU**, is a condition first indicated by the presence of phenylpyruvic acid in the urine. The continuous presence of glucose in the urine is an indication of **diabetes mellitus**. In addition the presence of acetone and other **ketone bodies** reveals faulty carbohydrate metabolism, a condition associated with **diabetes** and starvation. Huge amounts of the protein albumin in the urine may be an indication of renal (kidney) disorder.

The physical properties of urine may also provide information regarding a body malfunction. Normal freshly voided urine has an aromatic odor believed to be caused by the presence of rapidly evaporating acids. The urine of persons with bladder infections is usually foul-smelling because of the presence of bacteria. Persons with diabetes mellitus may have urine that has a fruity odor due to the presence of acetone.

The color of urine depends upon the concentration of its components, but it is usually amber because of the presence of a yellow pigment, urochrome. Excessive amounts of bile pigments will impart a dark yellow (almost orange) color to the urine, an indication of **obstructive jaundice**. The appearance of blood in the urine may be caused by kidney damage or by an infection of the urinary tract.

The normal specific gravity of urine is between 1.005 and 1.030. A continuously low reading may indicate **diabetes insipidus**, a disease caused by the impairment of the normal functioning of an antidiuretic hormone, whose function is to decrease the flow of urine in the kidney. Low specific gravity readings may also indicate that the kidney has lost its ability to concentrate the urine because of tubular damage.

High specific gravity readings occur in patients with diabetes mellitus, adrenal inefficiency, hepatic disease, or congestive cardiac failure. The specific gravity of urine is elevated because of the excessive loss of water caused by sweating, fever, vomiting, or diarrhea. Such extensive loss of water results in a high concentration of urinary constituents, particularly glucose and protein, and hence a high specific gravity.

The pH of normal urine can vary from 4.5 to 8.0. Persons on a normal diet excrete urine with a pH of approximately 6.0. Excessively acid urine, with a pH lower than 6.0, may be excreted by persons on a high protein diet. In addition, patients having acidosis or unchecked diabetes mellitus may excrete urine containing large amounts of acid. Normal urine is usually alkaline after meals, and individuals who are vegetarians usually void alkaline urine. Consistently high alkaline urine is an indication of a urinary tract infection.

Some substances not completely metabolized by the body are excreted unchanged in the urine. Quinine, nicotine, and excess vitamin C are typical examples.

Some of the methods you will employ in detecting the presence of inorganic ions (normal components of urine), glucose, and acetone are identical to those you have performed in previous experiments. The Keto-Diastix Reagent Strips will be used to detect both glucose and ketones. The Keto-Diastik Reagent Stix can be found through the following web sites: www.drugstore.com, www.americarx.com, and www. diabeticdrugstore.com. The strip is impregnated with the enzyme glucose oxidase to detect glucose and with nitroprusside for the detection of ketone bodies.

CAUTION: Clinitest tablets (Part IVC) contain sodium hydroxide. Avoid contact with skin, eyes, mucous membranes, and clothing.

Procedure

CAUTION: We require that disposable latex or vinyl gloves be worn during this experiment.

Use both commercial urine and a synthetic pathological sample supplied by your instructor in performing Parts I through IV.

I. PH

Use extended-range pH paper to check the pH of the urine sample. Record the pH. Note the color, odor, and cloudiness of the urine.

II. SPECIFIC GRAVITY

Use a urinometer to measure the specific gravity of the urine specimen. See Experiment 3, Part VI, for the proper procedure. Record the specific gravity.

III.INORGANIC COMPONENTS OF URINE

CAUTION: Nitric acid is caustic. After removing gloves, wash your hands.

A. Chloride

1. Pour 2 mL of urine into a test tube and add 1 mL of 6 M nitric acid. **CAUTION!** Wash your hands.

2. Add 1 mL of 0.1 M silver nitrate solution. A white precipitate indicates the presence of chloride in the urine.

3. Disposal: Discard the mixture into the "Waste Silver" container.

B. Sulfate

CAUTION: Hydrochloric acid is caustic. After removing gloves, wash your hands.

1. Pour 2 mL of urine into a test tube, and add 1 mL of 6 M HCl. **CAUTION!** Wash your hands.

2. Add 1 mL of 0.5 M $BaCl_2$. A white precipitate indicates the presence of sulfate.

3. Disposal: Discard the mixture into the "Waste Salts" container. Wash your hands.

C. Phosphate

CAUTION: This part of the experiment should be performed in a fume hood. Concentrated nitric acid is caustic. After removing gloves, wash your hands.

1. Pour 3 mL of urine into a test tube; then add about 3 drops of concentrated nitric acid (**CAUTION**) and 2 drops of 2.5% ammonium molybdate reagent. Wash your hands.

2. Place the test tube in a hot-water bath using a hot plate. The formation of a yellow-green precipitate (ammonium phosphomolybdate) indicates the presence of phosphate ion.

3. Disposal: Discard the mixture into the "Waste Salts" container.

D. Calcium

CAUTION: Potassium-oxalate is toxic. After removing gloves, wash your hands.

1. Pour 5 mL of urine into a test tube and add 5 drops of 20% potassium oxalate. A white precipitate indicates the presence of calcium ion.

2. Disposal: Discard the mixture into the "Waste Salts" container.

E. Ammonium Ion

CAUTION: Sodium hydroxide is caustic. After removing gloves, wash your hands.

1. Pour 5 mL of urine into a test tube and add 2 mL of 6 M sodium hydroxide. (**CAUTION!**) Wash your hands.

2. Gently heat the liquid in the test tube.

3. Place a strip of moist red litmus paper across the opening of the test tube. **Do not allow the litmus to touch the sides of the tube!**

4. Carefully smell the gas (NH_3) evolved.

5. Discard the contents of the test tube as indicated by the instructor.

IV. ORGANIC COMPONENTS OF URINE

CAUTION: Acetic acid is caustic. After removing gloves, wash your hands.

A. Albumin (a Protein)

1. Pour 5 mL of urine into a test tube and gently heat the upper portion of the tube.

2. If the urine becomes cloudy, add several drops of 5% acetic acid; then stopper and shake the test tube. If the cloudiness remains, albumin is present; if it disappears, the cloudiness was due to the presence of phosphates.

3. Discard the contents of the test tube as indicated by the instructor.

B. Glucose—Clinitest Reagent

CAUTION: Clinitest tablets contain sodium hydroxide. Avoid contact with skin, eyes, mucous membranes, and clothing.

1. Pour 1 mL of urine into a test tube and add 2 mL of deionized water.

2. Place the test tube in a rack and drop a Clinitest tablet into the solution. **(CAUTION!) Do not handle the test tube until 15 seconds after the reaction has stopped.** Wash your hands.

3. Remove the test tube from the rack, shake gently, and compare the color produced with the color chart.

4. Discard the contents of the test tube as indicated by the instructor.

C. Ketone Bodies—Legal Test

CAUTION: Glacial acetic acid and concentrated ammonium hydroxide are caustic. After removing gloves, wash your hands.

1. Pour 2 mL of urine into a clean test tube and add 2 drops of glacial acetic acid and 3 drops of 5% sodium nitroprusside solution. Wash your hands.

2. Incline the test tube and pour 1 mL of concentrated ammonium hydroxide (**CAUTION!**) down the side of the tube so that two layers are formed. Wash your hands. A red color at the boundary of the layers indicates the presence of acetone bodies.

3. Discard the contents of the test tube as indicated by the instructor.

D. Glucose and Acetone—Keto-Diastix Reagent

1. Pour 3 mL of urine into a test tube. Dip the reagent end of a Keto-Diastix strip in the specimen for **2 seconds** and remove it immediately.

2. Tap the strip against the sides of the test tube to remove the excess urine.

3. Compare the reagent side of the Keto-Diastix to the appropriate color chart. Take a ketone reading at **15 seconds** and a glucose reading after **30 seconds**. The time at which you take these readings is critical for obtaining accurate results.

4. Discard the test strip and the contents of the test tube as indicated by the instructor.

Name ______________________ Section __________ Date __________

Prelab Questions for Experiment 21

What physiological conditions are associated with the following:

a. Foul-smelling urine

____

____

b. Appearance of blood in the urine

____

____

c. Excessively high specific gravity readings

____

____

d. Continuously low specific gravity readings

____

____

e. Consistently alkaline urine

____

____

f. Deep yellow color in the urine

____

____

Name ______________________ Section ____________ Date ______________

Report for Experiment 21

I-III. URINE ANALYSIS—INORGANIC COMPONENTS

1. Data

Sample	*pH*	*Specific Gravity*	Cl^- (+ or −)	SO_4^{2-} (+ or −)	PO_4^{3-} (+ or −)	Ca^{2+} (+ or −)	NH_4^{1+} (+ or −)]

2. Describe the color, odor, and clarity of the urine tested.

____

____

____

3. a. What precipitate was formed when the urine was treated with silver nitrate?

____

b. Write an ionic equation to illustrate the precipitate's formation.

____

4. a. What precipitate was formed when the urine was treated with barium chloride?

____

b. Write an ionic equation to illustrate the precipitate's formation.

____

5. a. What precipitate was formed when the urine was treated with potassium oxalate?

____

b. Write an ionic equation to illustrate the precipitate's formation.

____

6. a. Explain your observations of the red litmus paper when you were testing for the presence of NH_4^+.

____

____

b. What odor did you detect while performing the above test?

____

IV. URINE ANALYSIS—ORGANIC COMPONENTS

1. Data:

Sample	*Albumin (+ or –)*	*Glucose*		*Ketone Bodies*	
		Clinitest (%)	*Diastix (%)*	*Legal (+ or –)*	*Diastix (%)*

2. Diagnose a possible physiological condition of the patient whose "pathological" urine you analyzed.

____

____

____

____

____

Name ______________________ Section __________ Date __________

Related Questions for Experiment 21

1. Explain how a urine analysis may help identify a drug addiction. (Hint: See Experiment 14.)

____

____

____

2. a. Does a positive glucose test always indicate a diabetic condition?

____

b. Explain your answer in question 2a.

____

____

____

3. What does the presence of ketone bodies indicate?

____

____

4. Give the names and structural formulas of ketone bodies that may be found in the urine.

5. What does a positive albumin test indicate?

____

____

6. a. Water-soluble vitamins in excessive doses (beyond the minimum requirements) are excreted in the urine. Will excessive doses of these vitamins likely be of benefit to an individual?

____

b. Explain your answer in a.

____

____

____

c. Discuss the possible harm that may be produced as a result of excessive intake of vitamins.

____

____

____

7. a. Write the structural formula of urea.

b. To what class of compound (alcohol, amine, etc.) does urea belong?

____

The Elements

Name	Symbol	Atomic number	Atomic weight[1]
Actinium	Ac	89	(227)
Aluminum	Al	13	26.98154
Americium	Am	95	(243)
Antimony	Sb	51	121.760
Argon	Ar	18	39.948
Arsenic	As	33	74.92160
Astatine	At	85	(210)
Barium	Ba	56	137.327
Berkelium	Bk	97	(247)
Beryllium	Be	4	9.012182
Bismuth	Bi	83	208.9804
Bohrium	Bh	107	(264)
Boron	B	5	10.811
Bromine	Br	35	79.904
Cadmium	Cd	48	112.411
Calcium	Ca	20	40.078
Californium	Cf	98	(251)
Carbon	C	6	12.0107
Cerium	Ce	58	140.116
Cesium	Cs	55	132.9054
Chlorine	Cl	17	35.453
Chromium	Cr	24	51.9961
Cobalt	Co	27	58.9332
Copper	Cu	29	63.546
Curium	Cm	96	(247)
Darmstadtium	Ds	110	(281)
Dubnium	Db	105	(262)
Dysprosium	Dy	66	162.50
Einsteinium	Es	99	(252)
Erbium	Er	68	167.259
Europium	Eu	63	151.964
Fermium	Fm	100	(257)
Fluorine	F	9	18.99840
Francium	Fr	87	(223)
Gadolinium	Gd	64	157.25
Gallium	Ga	31	69.723
Germanium	Ge	32	72.64
Gold	Au	79	196.9665
Hafnium	Hf	72	178.49
Hassium	Hs	108	(269)
Helium	He	2	4.00260
Holmium	Ho	67	164.93032
Hydrogen	H	1	1.00794
Indium	In	49	114.818
Iodine	I	53	126.9045
Iridium	Ir	77	192.217
Iron	Fe	26	55.845
Krypton	Kr	36	83.80
Lanthanum	La	57	138.9055
Lawrencium	Lr	103	(262)
Lead	Pb	82	207.2
Lithium	Li	3	6.941
Lutetium	Lu	71	174.967
Magnesium	Mg	12	24.3050
Manganese	Mn	25	54.93805
Meitnerium	Mt	109	(268)
Mendelevium	Md	101	(258)
Mercury	Hg	80	200.59
Molybdenum	Mo	42	95.94
Neodymium	Nd	60	144.24
Neon	Ne	10	20.1797
Neptunium	Np	93	(237)
Nickel	Ni	28	58.6934
Niobium	Nb	41	92.9063
Nitrogen	N	7	14.0067
Nobelium	No	102	(259)
Osmium	Os	76	190.23
Oxygen	O	8	15.9994
Palladium	Pd	46	106.42
Phosphorus	P	15	30.97376
Platinum	Pt	78	195.078
Plutonium	Pu	94	(244)
Polonium	Po	84	(209)
Potassium	K	19	39.0983
Praseodymium	Pr	59	140.90765
Promethium	Pm	61	(145)
Protactinium	Pa	91	231.0359
Radium	Ra	88	(226)
Radon	Rn	86	(222)
Rhenium	Re	75	186.207
Rhodium	Rh	45	102.9055
Roentgenium	Rg	111	(272)
Rubidium	Rb	37	85.4678
Ruthenium	Ru	44	101.07
Rutherfordium	Rf	104	(261)
Samarium	Sm	62	150.36
Scandium	Sc	21	44.95591
Seaborgium	Sg	106	(266)
Selenium	Se	34	78.96
Silicon	Si	14	28.0855
Silver	Ag	47	107.868
Sodium	Na	11	22.98977
Strontium	Sr	38	87.62
Sulfur	S	16	32.065
Tantalum	Ta	73	180.9497
Technetium	Tc	43	(98)
Tellurium	Te	52	127.60
Terbium	Tb	65	158.92534
Thallium	Tl	81	204.383
Thorium	Th	90	232.0381
Thulium	Tm	69	168.93421
Tin	Sn	50	118.710
Titanium	Ti	22	47.867
Tungsten	W	74	183.84
Ununbium[2]	Uub	112	(285)
Ununquadium[2]	Uuq	114	(289)
Uranium	U	92	238.0289
Vanadium	V	23	50.9415
Xenon	Xe	54	131.293
Ytterbium	Yb	70	173.04
Yttrium	Y	39	88.9058
Zinc	Zn	30	65.39
Zirconium	Zr	40	91.224

[1] For elements that do not have stable isotopes, the mass number of the most stable isotope is given in parentheses.
[2] Elements 112 and 116 have been reported, but have not been given official names. These elelments have been given temporary names, using the prefixes un (1), bi (2), and quad (4).

Periodic Table of the Elements

- Metals
- Semimetals
- Nonmetals

Representative elements: IA–IIA and IIIA–VIIIA. Transition metals: IIIB–IIB.

Period	IA	IIA	IIIB	IVB	VB	VIB	VIIB	VIIIB	VIIIB	VIIIB	IB	IIB	IIIA	IVA	VA	VIA	VIIA	VIIIA
1	1 **H** 1.00794																	2 **He** 4.00260
2	3 **Li** 6.941	4 **Be** 9.012182											5 **B** 10.811	6 **C** 12.0107	7 **N** 14.0067	8 **O** 15.9994	9 **F** 18.99840	10 **Ne** 20.1797
3	11 **Na** 22.98977	12 **Mg** 24.3050											13 **Al** 26.98154	14 **Si** 28.0855	15 **P** 30.97376	16 **S** 32.065	17 **Cl** 35.453	18 **Ar** 39.948
4	19 **K** 39.0983	20 **Ca** 40.078	21 **Sc** 44.95591	22 **Ti** 47.867	23 **V** 50.9415	24 **Cr** 51.9961	25 **Mn** 54.93805	26 **Fe** 55.845	27 **Co** 58.9332	28 **Ni** 58.6934	29 **Cu** 63.546	30 **Zn** 65.39	31 **Ga** 69.723	32 **Ge** 72.64	33 **As** 74.92160	34 **Se** 78.96	35 **Br** 79.904	36 **Kr** 83.80
5	37 **Rb** 85.4678	38 **Sr** 87.62	39 **Y** 88.9058	40 **Zr** 91.224	41 **Nb** 92.9063	42 **Mo** 95.94	43 **Tc** [98]	44 **Ru** 101.07	45 **Rh** 102.9055	46 **Pd** 106.42	47 **Ag** 107.868	48 **Cd** 112.411	49 **In** 114.818	50 **Sn** 118.710	51 **Sb** 121.760	52 **Te** 127.60	53 **I** 126.9045	54 **Xe** 131.293
6	55 **Cs** 132.9054	56 **Ba** 137.327	57 ***La** 138.9055	72 **Hf** 178.49	73 **Ta** 180.9479	74 **W** 183.84	75 **Re** 186.207	76 **Os** 190.23	77 **Ir** 192.217	78 **Pt** 195.078	79 **Au** 196.9665	80 **Hg** 200.59	81 **Tl** 204.383	82 **Pb** 207.2	83 **Bi** [208.9804]	84 **Po** [209]	85 **At** [210]	86 **Rn** [222]
7	87 **Fr** [223]	88 **Ra** [226]	89 **†Ac** [227]	104 **Rf** [261]	105 **Db** [262]	106 **Sg** [266]	107 **Bh** [264]	108 **Hs** [269]	109 **Mt** [268]	110 **Ds** [281]	111 **Rg** [272]	112 **Uub** [285]		114 **Uuq** [289]				

*Lanthanide elements	58 **Ce** 140.116	59 **Pr** 140.90765	60 **Nd** 144.24	61 **Pm** [145]	62 **Sm** 150.36	63 **Eu** 151.964	64 **Gd** 157.25	65 **Tb** 158.92534	66 **Dy** 162.50	67 **Ho** 164.93032	68 **Er** 167.259	69 **Tm** 168.93421	70 **Yb** 173.04	71 **Lu** 174.967
†Actinide series	90 **Th** 232.0381	91 **Pa** 231.0359	92 **U** 238.0289	93 **Np** [237]	94 **Pu** [244]	95 **Am** [243]	96 **Cm** [247]	97 **Bk** [247]	98 **Cf** [251]	99 **Es** [252]	100 **Fm** [257]	101 **Md** [258]	102 **No** [259]	103 **Lr** [262]

For elements that do not have stable isotopes, the mass of the most stable isotope is given in parentheses. Elements 112 and 114 have been reported but have not been given official names.

✍ Take Note!

✍ **Take Note!**

✍ Take Note!

✍ Take Note!

✍ Take Note!

✍ Take Note!

✍ Take Note!

✍ Take Note!

✍ Take Note!

✍ Take Note!

✍ Take Note!

✍ Take Note!

✍ **Take Note!**

✍ Take Note!

✍ Take Note!

✍ Take Note!

✍ Take Note!

✍ Take Note!

✍ Take Note!

✍ Take Note!

✍ Take Note!

✍ Take Note!

✍ Take Note!

✍ Take Note!

✍ Take Note!

✍ Take Note!

✍ Take Note!

✍ Take Note!

✍ Take Note!

✍ Take Note!